Building Micro Frontends with React 18

Develop and deploy scalable applications using
micro frontend strategies

Vinci J Rufus

BIRMINGHAM—MUMBAI

Building Micro Frontends with React 18

Copyright © 2023 Packt Publishing

Group Product Manager: Rohit Rajkumar

Publishing Product Manager: Kushal Dave

Senior Editor: Aamir Ahmed

Book Project Manager: Sonam Pandey

Technical Editor: K Bimala Singha

Copy Editor: Safis Editing

Proofreader: Safis Editing

Indexer: Manju Arasan

Production Designer: Ponraj Dhandapani

DevRel Marketing Coordinator: Nivedita Pandey

First published: September 2023

Production reference: 1180923

Published by Packt Publishing Ltd.

Grosvenor House

11 St Paul's Square

Birmingham

B3 1RB

ISBN 978-1-80461-096-1

www.packtpub.com

Contributors

About the author

Vinci J Rufus is VP of Technology with Publicis Sapient, with over 25 years of experience in leading teams that build rich web and internet applications for top retail brands. He was an early pioneer in adopting HTML5 and JavaScript frameworks such as Angular, React, and serverless technologies to deploy web apps in production.

For the past 5 years, Vinci has specialized in architecting and implementing micro frontends. He enjoys the technical challenges and complexity these systems entail.

At his core, Vinci is an engineer who loves collaborating directly with his team to solve real-world problems. He still codes regularly and stays hands-on with the latest technologies.

Vinci is particularly fascinated by emerging technologies such as edge computing and generative AI. He sees enormous potential in how these innovations can transform the way we develop software in the coming years.

Acknowledgments

To **Sheldon Monteiro** and the entire leadership team at the **Chief Marketing Technology Officer University (CMTOu)** – where my journey into micro frontends all began.

I am tremendously thankful for the guidance, support, and opportunities provided by my leaders at Publicis Sapient, **Tilak Doddapaneni** and **Rakesh Ravuri**. Your trust in allowing me to explore cutting-edge technologies has been instrumental in my knowledge of micro frontends. My sincere appreciation to the outstanding experience Engineering community at Publicis Sapient, from whom I have learned so much.

A huge thank you to **Sourav Mondal** for our many brainstorming and debugging. Thanks to **Nuala Monaghan**, **Devesh Kaushal**, **Hari Om Bangari** and **Davide Fortuna** for those additional reviews, comments and feedback. I am grateful for your time and insight. Most importantly, I wish to express my deepest love and gratitude to my amazing wife, *Raina*, and wonderful children, *Shannon* and *Jaden*. Your unrelenting support and understanding during our many sacrificed holidays have made this book possible. I could not have done this without you.

About the reviewers

Alberto Arias López is a continuously evolving software engineer with extensive work experience building consumer-focused online products and services in different domains within a number of international companies of all sizes. He has witnessed the evolution of the frontend, empowering developers to build more complex web applications each time that improve the user experience. Nevertheless, it is still a young software engineering area compared to others. Micro frontends is a step further that allows the community to keep evolving and responding to the increasing demand of the product life cycle.

The work I've done on this book is dedicated to my family: my parents, Ana María and Antonio, and my brothers, Antonio and Chumy. I love you.

I would also like to mention all the professional mates I have had the opportunity to know and learn from during the journey – teamwork is invaluable.

Israel Antonio Rosales Laguan is an experienced full stack software engineer using JavaScript, React, and Node.js, with a focus on process improvement, developer ergonomics, systems integration, and pipeline automation. He also has a lot of experience in international SCRUM teams and mentoring others, working in Equinox, OnDeck, and Lazard, among others. Other expertise: OWASP compliance, GraphQL, CI/CD with Docker, and advanced CSS.

Table of Contents

3

Part 2: Architecting Microfrontends

4

5

6

Server-Rendered Microfrontends   99

Part 3: Deploying Microfrontends

7

Deploying Microfrontends to Static Storage   121

8

Deploying Microfrontends to Kubernetes   137

Part 4: Managing Microfrontends

9

Managing Microfrontends in Production   165

10

Common Pitfalls to avoid when Building Microfrontends   173

Part 5: Emerging Trends

11

Preface

Microfrontends have emerged as a popular architectural pattern for building large, complex web applications by breaking them down into smaller, independent pieces that can be developed and deployed separately. This makes it easier to scale development, accelerate release velocity, and adopt new technologies incrementally.

This book provides a comprehensive guide to implementing micro frontends in practice, from architectural principles and patterns to hands-on examples using frameworks such as React. You will learn how to divide a monolithic application into autonomous micro frontends. We will look at some of the key principles of micro frontends and scenarios where micro frontends may or may not be the right pattern.

We will explore the different patterns of micro frontends and create micro frontends using module federation for client-side rendered apps and server-side rendered apps.

We will learn how to deal with things such as routing and state management when building micro frontends, and finally, we will learn how to deploy our micro frontends on Firebase and on a Kubernetes cluster using Azure.

Who this book is for

This book is for frontend and full stack developers aiming to build large, scalable web applications using modern JavaScript frameworks such as React. It will also benefit solution architects looking to adopt micro frontend architecture. You should have a good understanding of JavaScript, React, module bundling, and basic web development concepts.

What this book covers

Chapter 1, *Introducing Microfrontends*, introduces different architectural patterns, such as the multi-SPA and micro apps pattern for building micro frontends.

Chapter 2, *Key Principles and Components of Microfrontends*, covers core principles such as independent deployability, bounded contexts, isolating failures, runtime integrations, and so on.

Chapter 3, *Monorepos versus Polyrepos for Microfrontends*, compares monorepos and multirepos for managing micro frontend code bases, and why monorepos are preferred for building microfrontends.

Chapter 4, *Implementing the Multi-SPA Pattern for Microfrontends*, demonstrates building micro frontends as a collection of Single-Page Apps.

Chapter 5, Implementing the Micro-Apps Pattern for Microfrontends, dives deeper into building micro frontends using module federation and covering critical topics around routing and sharing state between different micro apps.

Chapter 6, Server-Rendered Microfrontends, shows how to go about building a server-side rendered micro-frontend using module federation.

Chapter 7, Deploying Microfrontends to Static Storage, takes us through the journey of deploying our micro frontend to a static storage hosting service such as Firebase.

Chapter 8, Deploying Microfrontends to Kubernetes, demonstrates deploying micro frontends to Kubernetes such as AKS on Azure.

Chapter 9, Managing Microfrontends in Production, covers topics such as branching strategies, versioning, rollback strategies, and feature toggles that are essential to managing micro frontends in production.

Chapter 10, Common Pitfalls to avoid when Building Microfrontends, talks about some of the common mistakes developers and architects make that negatively impact the benefits of why we chose micro frontends in the first place.

Chapter 11, Latest Trends in Microfrontends, covers some of the new trends, such as ES builds, cloud or edge functions, island patterns, and generative AI, and how they could be used to build micro frontends in the future.

To get the most out of this book

The code examples use React, webpack, Node.js, and npm. Familiarity with these tools will be helpful. The examples can be followed on any operating system.

The digital version of this book includes detailed code examples that can be copied and pasted to get up and running quickly. For the best learning experience, try building the examples yourself from scratch.

Software/hardware covered in the book	Operating system requirements
React 18	Windows, macOS, or Linux
TypeScript 3.7	Windows, macOS, or Linux
Docker Engine 24	Windows, macOS, or Linux

If you are using the digital version of this book, we advise you to type the code yourself or access the code from the book's GitHub repository (a link is available in the next section). Doing so will help you avoid any potential errors related to the copying and pasting of code.

Download the example code files

You can download the example code files for this book from GitHub at https://github.com/PacktPublishing/Building-Micro-Frontends-with-React-18. If there's an update to the code, it will be updated in the GitHub repository.

We also have other code bundles from our rich catalog of books and videos available at https://github.com/PacktPublishing/. Check them out!

Conventions used

There are a number of text conventions used throughout this book.

`Code in text`: Indicates code words in text, database table names, folder names, filenames, file extensions, pathnames, dummy URLs, user input, and Twitter handles. Here is an example: "Ensure we have the URL routing set up in the `proxy.conf.json` file."

A block of code is set as follows:

```
"scripts": {
  "start": "nx serve",
  "build": "nx build",
  "test": "nx test",
  "serve:all": "nx run-many --target=serve"
},
```

When we wish to draw your attention to a particular part of a code block, the relevant lines or items are set in bold:

```
pnpm install semantic-ui-react semantic-ui-css
```

Any command-line input or output is written as follows:

```
pnpm serve:all
```

Bold: Indicates a new term, an important word, or words that you see onscreen. For instance, words in menus or dialog boxes appear in **bold**. Here is an example: "Next, we update the **Add** and **Remove** button onclick events as follows."

> **Tips or important notes**
> Appear like this.

Get in touch

Feedback from our readers is always welcome.

General feedback: If you have questions about any aspect of this book, email us at customercare@packtpub.com and mention the book title in the subject of your message.

Errata: Although we have taken every care to ensure the accuracy of our content, mistakes do happen. If you have found a mistake in this book, we would be grateful if you would report this to us. Please visit www.packtpub.com/support/errata and fill in the form.

Piracy: If you come across any illegal copies of our works in any form on the internet, we would be grateful if you would provide us with the location address or website name. Please contact us at copyright@packt.com with a link to the material.

If you are interested in becoming an author: If there is a topic that you have expertise in and you are interested in either writing or contributing to a book, please visit authors.packtpub.com

Share Your Thoughts

Once you've read *Building Micro Frontends with React 18*, we'd love to hear your thoughts! Scan the QR code below to go straight to the Amazon review page for this book and share your feedback.

https://packt.link/r/1-804-61096-8

Your review is important to us and the tech community and will help us make sure we're delivering excellent quality content.

Download a free PDF copy of this book

Thanks for purchasing this book!

Do you like to read on the go but are unable to carry your print books everywhere?

Is your eBook purchase not compatible with the device of your choice?

Don't worry, now with every Packt book you get a DRM-free PDF version of that book at no cost.

Read anywhere, any place, on any device. Search, copy, and paste code from your favorite technical books directly into your application.

The perks don't stop there, you can get exclusive access to discounts, newsletters, and great free content in your inbox daily

Follow these simple steps to get the benefits:

1. Scan the QR code or visit the link below

https://packt.link/free-ebook/9781804610961

2. Submit your proof of purchase
3. That's it! We'll send your free PDF and other benefits to your email directly

Part 1:
Introduction to Microfrontends

This part covers the core concepts and principles behind microfrontends, including the motivations for using this architecture, key components, and how microfrontends differ from monoliths.

This part has the following chapters:

- *Chapter 1, Introducing Microfrontends*
- *Chapter 2, Key Principles and Components of Microfrontends*
- *Chapter 3, Monorepos versus Polyrepos for Microfrontends*

Introducing Microfrontends

We are coming full circle with microfrontends! During the Web 1.0 era, websites primarily comprised single pages built in ASP, JSP, or PHP, where we could make changes to each individual page and upload it to a server via FTP and it was immediately available to consumers. Then came the Web 2.0 era and the notion of web apps and **Single-Page Apps (SPAs)**, where we compile, transpile, and deploy large monolithic apps. Now, we seem to be going back to working with smaller apps and pages.

The early 2000s brought in the era of Web 2.0 and the notion of web apps. A few years later, JavaScript frameworks allowed you to build SPAs that updated instantly and didn't reload a new page each time the user clicked on a link or a button. SPAs were indeed fast for small to medium-sized apps, but as teams went full throttle with building large-scale SPAs, and as applications and teams grew, the velocity and speed of development dropped significantly. Teams seemed to be debating about folder structures, state management, and breaking each other's code, due to centrally managed libraries and so on. These large SPAs also started becoming less performant due to the large bundle sizes of these apps. More importantly, the high execution time required to parse these JavaScript bundles made the apps even more sluggish on low-end devices and mobile phones. That's when developers and architects started looking for solutions to these problems. Thankfully, they didn't have to look too far.

You see, the backend teams went through the exact same problems with the large backend monoliths a few decades back and moved toward the microservices architecture pattern in order to solve their performance and scaling challenges. The frontend teams now look to apply the same principles of microservices to their frontend apps, which are being referred to as **microfrontends**.

The journey for backend teams toward microservices has been a very long one, spanning multiple decades, and many teams still struggle with it. However, thanks to a lot of debates, discussions, thoughts, leadership, and sharing learning from various microservice implementations, there is an overall maturity to and consensus around microservices architecture.

Frontend teams are just waking up to the notion of microfrontends, and there are multiple schools of thought on what defines a microfrontend, including, in fact, whether microfrontends are even a good thing or not. It will take a couple of years, if not a decade, before there is some consensus around microfrontends. The good thing, however, is that we can learn a lot from the journey of microservices, as a lot of principles and architecture patterns of microservices also apply to microfrontends.

In this chapter, we'll start by understanding the need for microfrontends. We will cover the definition of microfrontends, and then the different patterns of microfrontends. We will also look into the parameters that will help us choose which pattern to go with for designing your apps. Finally, we will create our very first microfrontend.

In this chapter, we will cover the following topics:

- Defining Microfrontends
- Understanding Microfrontend patterns
- Choosing a suitable pattern
- Hello World with Microfrontends

By the end of this chapter, you will have a better understanding of two of the most common patterns for building microfrontends and a guide to help you decide which one would be most suitable for you.

Toward the end of this chapter, we will build out a simple multi-SPA microfrontend example and get a feel for how we navigate between the the different SPAs.

Technical requirements

As you go through the code examples in this chapter, you will need the following:

- A PC, Mac, or Linux desktop/laptop with at least 8 GB of RAM (16 GB preferred)
- An Intel chipset i5+, AMD, or an Apple M1 + chipset
- At least 256 GB of free hard disk storage

You will also need the following software installed on your computer:

- Node.js version 16+ (use `nvm` to manage different versions of Node.js if you have to).
- Terminal: A modern shell such as `zsh`, iTerm2 with `oh-my-zsh` for Mac (you will thank me later), or Hyper for Windows (`https://hyper.is/`).
- IDE: We recommend VS Code.
- `npm`, `yarn`, or pnpm. We recommend PNPM because it's fast and storage efficient.
- Browser: Chrome/Microsoft Edge, Brave, or Firefox (I use Firefox).

The code files for this chapter can be found here: `https://github.com/PacktPublishing/Building-Micro-Frontends-with-React`.

Defining Microfrontends

In this section, we will focus on defining what microfrontends and their key benefits are, and also become aware of the initial upfront investments associated with setting up microfrontends.

The currently accepted definition of a microfrontend is as follows.

"Microfrontends are a composition of micro apps that can be **independently deployed** and are owned by **independent teams** responsible for delivering business value of a focused area of the overall application".

The keywords in this definition are independently deployed and independent teams. If at least one of these terms doesn't apply to you or your team, then you probably don't need a microfrontend. A regular SPA would work out to be more efficient and productive. As we will see later, microfrontends come with a bit of upfront complexity and may not be worth it unless you have a large application, where sections of the app are managed by individual teams.

We've noticed that some teams that are on their journey to implementing microfrontends misinterpret the *micro* part of microfrontends and believe an application doesn't follow a microfrontend architecture unless it's broken down to its smallest level. They break down their apps into really small apps, which adds a lot of unnecessary complexity. In fact, it negates all the benefits that microfrontends are supposed to deliver.

In our opinion, it actually works the other way around. When breaking down an application into micro apps, the teams should ideally look to identify the largest possible micro app or micro apps that a scrum team can independently manage and deploy to production without impacting other micro apps.

The key takeaway from this is not to be swayed by the term "micro" but instead identify the largest possible app that can be independently deployed by a single scrum team.

Before we go deeper into the wonderful world of microfrontends, it is important to remember that every application doesn't need to be a microfrontend. Let's learn more about this in the following section.

Understanding the Microfrontend Premium

Martin Fowler talks about the microservice premium. This refers to the fact that microservices come with a bit of overhead and complexity, mainly in terms of the initial setup and the communication channels between the services. Martin goes on to say that the benefits of a microservices architecture only start showing when size and complexity boosters kick in. To understand this, let's look at the following diagram:

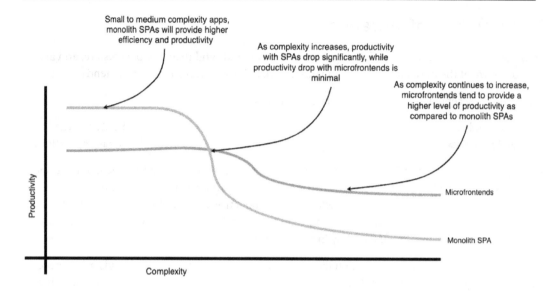

Figure 1.1 – The microservice premium graph (source: https://
martinfowler.com/bliki/MicroservicePremium.html)

The preceding diagram is a graph of the productivity versus the complexity of an application and depicts the drop in productivity for a monolith SPA and microfrontend as complexity grows.

The same holds true for the microfrontend architecture. The whole process of decoupling the various parts of components, routing, and templates and delegating them to different systems can become an unnecessary overhead for small or medium-scale apps.

The benefits of microfrontends kick in only when your project starts reaching the size and complexity thresholds shown in *Figure 1.1*.

Exploring the benefits of Microfrontends

All the benefits of a microfrontend architecture are linked to size and scale. Having said that, the following benefits of microfrontends hold true only for apps that are built and supported by teams with over 15 people.

In the following sections, we will learn about the benefits that teams can expect when they implement a microfrontend architecture, all of which are directly linked to improved productivity and better developer experience for team members.

Faster development and deployments

One of the main drawbacks of monolithic Single Page Apps is that as the application and team sizes grow, feature development and deployments come to a crawl. We notice the team spending a lot more time where one team is waiting on the other team to finish something before the application can be deployed. With a microfrontend architecture, every scrum team works independently on their micro app, building and releasing features without having to worry a lot about what other teams are doing.

Easier to scale as the application grows

A microfrontend architecture is all about composing smaller micro apps, so as the application grows in size, it's just a question of adding additional micro apps and having a scrum team own it.

Now, since each team deals with a smaller micro app, their team members need to spend less time understanding the code base and should not get overwhelmed or worried about how their code changes will impact other teams.

Microfrontends allow one to scale up very quickly, with scrum teams working in parallel once the base microfrontend framework is set up.

Improved Developer Experience

With isolated, independent micro apps, the time required for each team to compile, build, and run automated unit tests for their part of the micro apps is greatly reduced. This allows teams to build and deliver features a lot faster.

While teams run isolated unit and automation tests for their micro apps more frequently, we recommend running full regression suites of end-to-end tests on demand or before committing the code to Git.

Progressive upgrades

The frontend ecosystem is the fastest-evolving ecosystem. Every few months, a new framework or library springs up that is better and faster than the previous one. Having said that, there is always an urge to rewrite your existing application using the latest framework.

With large applications, it's not possible to easily upgrade or introduce a new framework without rewriting the entire application. The cost of rewriting the application and the associated risks of introducing bugs due to the rewrite are far too high. Teams keep deprioritizing the upgrade and within a few years, they find themselves working on an outdated framework.

With microfrontends, it is easier to pick up one small micro app and upgrade it or rewrite it and then gradually roll it out to other micro apps. This also allows teams to experience the benefits of the new change and learn and course-correct as they migrate the new framework to the other micro apps.

As we move on to the next section, let's quickly recap some of the key points that we've learned so far:

- Microfrontends are suited for building large-scale apps where teams are set up as full-stack teams, where the backend developers, frontend developers, product owners, and so on are within the same scrum team.

- Microfrontends have numerous benefits, such as team independence, features launched with improved velocity, and better developer experience. However, these benefits will start becoming visible once you have overcome the initial phase of complexity associated with the "microfrontend premium."

Understanding Microfrontend patterns

When it comes to microfrontends, there are way too many interpretations. These are still early days for microfrontends, and there is no right or wrong way of building them. The answer to any technical/architectural question is "It depends…." In this section, we will focus on two of the most common patterns that teams adopt while building microfrontends. We will see what key factors to consider when deciding which pattern may be right for you. We will end this section by building a really basic microfrontend to get the ball rolling.

At a very high level, there are two primary patterns for microfrontends. Both of these patterns can be applied irrespective of whether you are building a **Server-Side-Rendered** (**SSR**) app or a **Client-Side-Rendered** (**CSR**) app. To better illustrate these patterns, we will take the use case of an e-commerce application such as Amazon.

In the following subsections, we will look at these two patterns and how they differ from each other.

The Multi-SPA Pattern

The first pattern that we will discuss is the **multi-SPA** pattern. As the name suggests, the application is built up of multiple SPAs. Here, the app is broken down into 2-3 distinct SPAs and each app is rendered at its own URL. When the user navigates from one SPA to another, they are redirected via a browser reload. In the case of an e-commerce application, we could look at the search, product listing, and product details as one SPA, and the cart and checkout as the other SPA. Similarly, the **My Accounts** section, which includes the login, registration, and profile information, would form the third SPA.

The following figure shows an illustration of a multi-SPA pattern microfrontend for an e-commerce app:

mysite.com

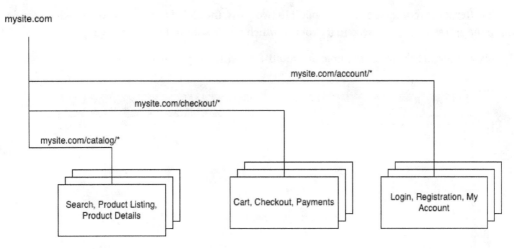

Figure 1.2 – Multi-SPA pattern microfrontend for an e-commerce app

As you can see from the preceding figure, our e-commerce application consists of three SPAs: the Catalog SPA, the Checkout SPA, and the Accounts SPA.

In the simplest form of this pattern, each app behaves as an independent SPA that sits within its own unique global URL.

Each SPA is deployed at a unique global route. For example, the catalog app would be deployed at a URL such as `mysite.com/catalog/*` and all subsequent secondary routes within the catalog app will load up as an SPA within the `/catalog/*` route.

Similarly, the accounts app would live in the global route of `mysite.com/accounts/` and the different pages within the account's app login, signup, and profile would be available at URLs such as `mysite.com/accounts/login` or `mysite.com/accounts/register`.

As mentioned earlier, when the user moves from one macro app to another, there will be a reload of the page in the browser. This is because we usually use the HTML `href` tags to navigate between the apps. This browser refresh is perfectly fine. I've seen teams go to great lengths, complicating their architecture, to try to achieve a single-page experience. The truth, however, is that users don't really care if your app is an SPA or a **Multi-Page App** (**MPA**). As long as the experience is fast and non-janky, they are happy.

At times, the browser reload may work in your favor as it will reduce the risks of memory bloat due to either memory leaks or too much data being put into a data store.

However, if you really want to nail that SPA experience, then you can always create a thin app shell that hosts the global routes and data store, such that each app is called within this app shell. *We will be going into more detail of this pattern in the upcoming chapters.*

In this pattern, the routing is generally split into two parts, the global or primary routes, which reside within the app shell, and the secondary routes, which reside within the respective apps.

The following figure shows an example of a multi-SPA with an app shell:

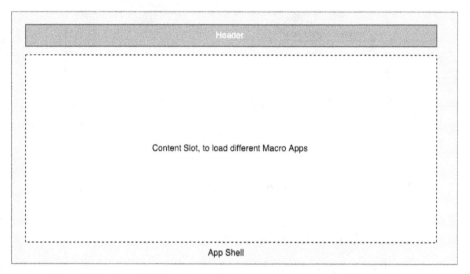

Figure 1.3 – A multi-SPA pattern with an app shell to give an SPA experience

Here, you will notice that we have introduced the notion of an app shell, which incorporates the header component, and the different SPAs load within the content slot. This pattern gives a true SPA experience as the header component doesn't refresh when transitioning from one SPA to the other.

The Micro Apps Pattern

The other pattern for building microfrontends is what we call the **micro apps** pattern. The reason we call it the micro apps pattern is that this is a more granular breakdown of the application.

As you can see in *Figure 1.4*, the web page is composed of different components where each component is an independent micro app that can exist in isolation and work in tandem with other micro apps as part of the same page.

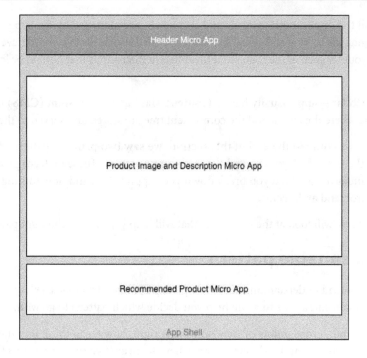

Figure 1.4 – Micro app architecture with product images and
recommended products co-existing as different micro apps

You will notice the preceding diagram is a more granular version of *Figure 1.3*, where we further break down the central content slot into smaller micro apps. Notice how the central content area now consists of two micro apps, namely the product details and recommended products micro apps.

The micro apps pattern is a lot more complex than the multi-SPA pattern and it is recommended mainly for very large web applications, where there are multiple teams that own different elements on a single page.

In *Figure 1.4*, we would assume that there is a dedicated team that manages the product description component of the page, and another team that manages the product recommendations component on the same page.

We would also assume that the frequencies at which these components get updated with feature enhancements would be different; for example, the recommendations micro app would constantly undergo A/B tests, and hence would need to be deployed more frequently than the product image and description micro app, which may not change as often.

In this pattern, all the routes, both primary and secondary, are managed by the app shell. Here, in addition to managing the routing and global states, the app shell also needs to store/retrieve information about the page layout for each of the routes and the different micro apps that need to be loaded within each of the pages.

In most cases, such large apps usually have a **Content Management System (CMS)** in place or a templating engine where the layout and the component tree are stored and served to the frontend.

To summarize, as we come to the end of this section, we saw two primary patterns for building microfrontends, the multi-SPA pattern and the micro apps pattern. These patterns primarily differ in the level of granularity at which you break down your application, and how routing is managed within the microfrontend architecture.

In the next section, we will look at the guidelines that will help you choose the right pattern.

Choosing a suitable pattern

Now that we have a broad understanding of the two patterns of microfrontends, let's spend some time on some of the key considerations that will help you decide which pattern to go with.

While there may be numerous points of view on what is right, how far to think into the future, and how to future-proof your app and architecture, we believe there are two primary factors that will help you decide on which of the two patterns to go with for your microfrontend architecture. Let's look at them in detail in the following sections.

Team Composition

For teams that build applications on microservices and microfrontends, it is a common practice that they are vertically sliced based on business functionality. In the e-commerce example, we may have a team that focuses on the browsing journey and another team that focuses on the checkout journey. If one scrum team owns the entire browser journey and one scrum team owns the entire checkout journey, then it is recommended that you go for the multi-SPA pattern. However, if you have numerous small teams that own different entities of the business domain, such as, say, search, product recommendations, and promotions, then it would be wise to go for the micro apps pattern. As mentioned earlier, the rule of thumb is for each scrum team to ideally own a single micro app.

Frequency of Deployments

Another factor that would come into play when deciding how to break down your microfrontend would be the frequency of deployments. If there are specific sections of the app that change more than others, then those sections can be separated into its own microfrontend, which can be separately deployed without affecting the other sections of the app. This reduces the amount of testing that needs to be done because now we need to test only the micro app that is being changed and not the entire application.

As we can see, the decision on whether you should go for a multi-SPA pattern or the micro apps pattern boils down to the two key factors of team composition and deployment frequency, and this is directly related to the two keywords from the definition of microfrontend, namely, independent teams and independent deployments.

Hello World with Microfrontends

OK, it's time to get our hands dirty writing some code. We are going to start simple by building a basic multi-SPA pattern app. In this example, we will use Next.js, which is currently the most popular tool for building performant React applications. Follow these steps:

> **Note**
>
> For the rest of this chapter, we assume you are using pnpm as the package manager. If not, replace pnpm with npm in the respective commands.

1. Let's start by creating a root folder for our app. We'll call it my-store. Run the following command in your terminal:

   ```
   mkdir my-store
   ```

2. Now, let's cd into my-store and create our two Next.js apps, namely, home and catalog, by typing the following commands in our terminal:

   ```
   cd my-store
   pnpm create-next-app@12
   ```

   Or, we can type the following:

   ```
   cd my-store
   npx create-next-app@12
   ```

3. When it prompts you to add a project name, call it home. It will then go through the various steps and complete the installation.

   The interesting thing about create-next-app is even through you define the version as @12, it will nevertheless pull the latest version of Next.js, hence to ensure consistency with the rest of this chapter we will update the version of next in package.json as follows:

   ```
   "dependencies": {
       "next": "12",
       "react": "18.2.0",
       "react-dom": "18.2.0"
   ```

4. Now delete the node_modules folder and the package lock file and run the pnpm i command

> **Important note**
>
> While you can always use yarn or npx to run the CLI, we recommend using pnpm as it is 2-3 times faster than npm or yarn.

5. Once it's done with the setup, go ahead and create another app repeating steps 2-5. Let's call this project catalog.

   Once complete, your folder structure would look as follows:

```
└── my-store/
    ├── home
    └── catalog
```

6. Now, let's run the home app by typing the following commands:

```
cd home
pnpm run dev
```

7. Your app should now be served on port 3000. Verify it by visiting http://localhost:3000 on your browser.

8. Let's get rid of the boilerplate code and add simple navigation. Locate and open up the file located at home/pages/index.js and replace everything within the <main></main> tags with the following:

```
<main className={styles.main}>
  <nav><a href="/">Home</a> | <a href="/catalog">Catalog</a> </nav>
    <h1 className={styles.title}>
      Home:Hello World!
    </h1>
    <h2>Welcome to my store</h2>
</main>
```

Note that we've added basic navigation to navigate between the home and catalog pages. Your home app that is running on `localhost:3000` should now look as follows:

Figure 1.5 – Screenshot of the home app with two navigation links for Home and Catalog

9. Now, let's move on to the catalog app. Navigate to the index page, located at `/catalog/pages/index.js`, and again, let's get rid of the boilerplate code and replace the contents within the `<main>` tag with the following code:

```
<main className={styles.main}>
    <nav><a href="/">Home</a> | <a href="/catalog">Catalog</a> </nav>
    <h1 className={styles.title}>
        Catalog:Hello World!
    </h1>
    <h2>List of Products</h2>
</main>
```

Now, since we already have the home page being served on port 3000, we will run our catalog app on port 3001.

10. We do this by adding the port flag for the `dev` command within the `scripts` section of the `catalog/package.json` file, as follows:

```
"scripts": {
    "dev": "next dev -p 3001
    ...
}
```

11. Now, running `pnpm run dev` from within the catalog app should run the catalog app on `http://localhost:3001`. You can see this in the following screenshot:

Figure 1.6 – Screenshot of the catalog app running on port 3001

The next step is to wire these up such that when the user hits `localhost:3000`, it directs them to the home app, and when the user hits `localhost:3000/catalog`, they are redirected to the catalog app. This is to ensure that both apps feel as if they are part of the same app, even though they are running on different ports.

12. We do this by setting the `rewrites` rule in the `home/next.config.js` file, as follows:

```
const nextConfig = {
  reactStrictMode: true,
  swcMinify: true,
  async rewrites() {
    return [
      {
        source: '/:path*',
        destination: `/:path*`,
      },
      {
        source: '/catalog',
        destination: `http://localhost:3001/catalog`,
      },
      {
        source: '/catalog/:path*',
        destination: `http://localhost:3001/catalog/:path*`,
      },
    ]
  },
}

module.exports = nextConfig
```

As you can see from the preceding code, we simply tell Next.js that if the source URL is `/catalog`, then load the app from `localhost:3001/catalog`.

13. Before we test it out, there is another small change needed to the catalog app. As you can see, the catalog app will be served on the root of port `3001`, but what we would like is for it to be served at `:3000/catalog`. This is because with the rewrite we did earlier, Next.js will expect the catalog apps and its assets to be available at `/catalog/*`. We can do this by setting the `basePath` variable in the `catalog/next.config.js` file as follows:

```
const nextConfig = {
  reactStrictMode: true,
  swcMinify: true,
  basePath:'/catalog'
}
```

14. Now, to test that this is working fine, we will run up both of the apps in two different terminal windows by navigating to the home and catalog apps and running the `pnpm run dev` command.

15. Open up `http://localhost:3000` in your browser and verify that the home app is loaded. Click on the **Catalog** link and verify that the catalog page does load up at `http://localhost:3000/catalog`. Notice that the app catalog that's running individually on port `3001` is sort of "proxied" to load up within a unique URL of the parent/host app. This is one of the key principles of microfrontends, where apps running on different ports and different locations are "stitched" together to make it look like they are a part of the same application.

With that, we come to the end of creating our very first microfrontend with the multi-SPA pattern. We will look at the micro apps pattern in more detail in the upcoming chapters. This pattern meets the majority of the use cases for building microfrontends and checks all the key principles of microfrontends, which we are going to see in the next chapter.

Summary

It's a wrap for this chapter. We started off by learning how microfrontends (when executed correctly) help teams to continue to release new features at a consistent pace even as the app size and complexity grow. Then, we learned that there are two primary patterns for implementing microfrontends, the multi-SPA pattern and the micro apps pattern. We saw that the multi-SPA pattern is easier to implement and would suit the majority of use cases. The micro apps pattern would be more suitable when different elements of a given page are owned by different scrum teams. Finally, we learned how to build our very own microfrontend application and saw how we can navigate between the two apps while still giving the user the illusion that they are both part of a single app.

In the next chapter, we will look at some of the key principles to strictly adhere to when designing your microfrontend architecture. We will also look at some of the key components of microfrontend and the various ways they can be implemented.

2
Key Principles and Components of Microfrontends

Microfrontends are a double-edged sword. When done right, they can bring a great amount of joy and productivity to teams; however, if not implemented the right way, they can make things way worse.

Having said that, there are a couple of key principles and considerations we need to keep in mind when building a microfrontend architecture.

In this chapter, we will look at the key design principles of a microfrontend architecture and why it is important to treat them as sacrosanct. The reason we emphasize these principles is that they lay the foundation of the microfrontend architecture. Teams may not be able to extract all the benefits of a microfrontend pattern if they choose to ignore these principles. Then, we will look at the key components that are critical to any microfrontend architecture.

In this chapter, we will cover the following topics:

- Understanding the Key Principles
- The key Components of a Microfrontend Architecture

By the end of this chapter, you will have a better understanding of the guiding principles and key considerations that teams need to keep in mind when designing a microfrontend architecture.

Understanding the Key Principles

It's important that all software teams lay down a set of rules and guiding principles that all team members and the code they write adhere to. This ensures that when teams discuss certain technical approaches, they can validate them against these guidelines. This, in turn, ensures that the teams can focus on the outcomes by mapping them against these key guidelines and not get too obsessed with the nuances of the process. This helps teams to arrive at decisions a lot faster.

In the following sections, we will look at the key principles that teams must adhere to when following a microfrontend pattern.

Domain Driven Teams

Dan Abramov, who leads the React project at Meta, once tweeted the question, *"Is Microfrontends solving a technology problem or an organizational problem?"*

When you think about it, a lot of problems we see in today's software development do stem from the way teams are organized.

Domain Driven Design is a well-established concept in the microservices world. Backend microservice teams are commonly organized around these domain models. With microfrontends, we extend the same thinking to the frontend world, and by re-organizing the frontend teams within these domain models, we are now able to create vertically sliced teams, where a domain-driven team can own the responsibility of a business functionality from end to end and is able to work independently.

For us to be successful with microfrontends, it is critical that the micro apps and teams that own them are mapped to these domain models and the business value they aim to provide.

Let's have a quick look at what a Domain Driven Team might look like:

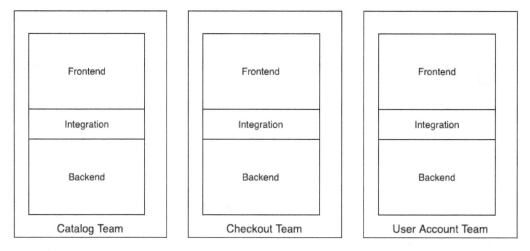

Figure 2.1 – Domain-driven teams

The preceding diagram shows three domain-driven teams for an e-commerce application, namely **Catalog Team**, **Checkout Team**, and **User Account Team**. Within each team, you will see that they have dedicated team members who play the roles of frontend, backend, and integration engineers.

Isolating Failure

Microfrontends are inherently designed to be "decentralized." One of the many benefits of that is isolating failures and reducing the blast radius of an error. A common problem with monolith **Single Page Apps** (**SPAs**) is that a single line of error in any one of the modules would prevent an entire application from being compiled, or a runtime error would cause an entire page to error out.

When designing a microfrontend architecture, you need to ensure graceful service degradation if one or more of the microfrontends fail.

If one microfrontend is dependent on another for its functioning, then we are breaking one of the key principles of microfrontends, which should be avoided at all costs.

Deploying Independently

Another key principle of a microfrontend architecture is the ability to deploy each app independently without having to redeploy the other apps.

When a new app is deployed, it should immediately be available to a user and should not require a restart of the host app or the servers for the changes to take effect.

An interesting observation with different teams working on microfrontends is that while, from an architecture standpoint, these micro apps can be updated independently, the DevOps pipelines that deploy these microfrontends are designed to deploy all the apps simultaneously, thereby negating the benefits of independent deployment.

It is critical that the DevOps pipelines are also designed such that when any app is ready for deployment, only the relevant pipeline runs and deploys the app, without impacting the other apps.

This misconfiguration of the DevOps pipelines mainly stems from the problem where there are separate DevOps teams that are responsible for building the pipelines and production deployments.

The best way to fix this is to ensure that we have "full life cycle teams," who are responsible for building the app and also responsible for deploying it to production. These teams work closely with the DevOps teams to build the CI and CD pipelines and then take over the control of managing and running them.

Preferring Runtime Integrations

A common discussion in the context of microfrontends is build time integrations versus runtime integrations. With build time integrations, the different teams build and publish their micro apps either to a version control system or an artifact repository, such as NPM or Nexus.

Then, during building time, all these micro apps are brought together to build a single app bundle, which is then deployed to production. We strongly discourage this pattern of build-time integration, as it breaks the aforementioned principle of independent deployment. A pattern like this may be suitable where you have scheduled releases that happen either once or twice a month. However, in that case, you probably would be better off with a monolith single-page app and don't really have to deal with all the complexities of a microfrontend architecture.

Always prefer runtime integrations when designing a microfrontend architecture.

Your micro apps should immediately be available for use the moment they are deployed. This ensures that each team can continuously deploy their micro apps to production and are not dependent on other teams to make their app available.

In most microfrontend patterns, we can make use of a host application or a shell app that keeps a tab of the different micro apps that load within it, but care must be taken to ensure that this host/app shell is built with scalability in mind. If the process of checking for new versions of a micro app takes up a lot of CPU or memory resources, then there is a high risk that it will become a single point of failure when your application scales, in terms of the micro apps and the traffic it receives.

Avoiding the "Distributed Monolith" trap

Don't Repeat Yourself (DRY) has a slightly different meaning in the microservice/microfrontend world. Most developers associate DRY with code reusability. When working with microfrontends, teams can go overboard creating libraries and utilities, which eventually get imported and used in each of the micro apps. Now, as each team's needs grow, they start adding functionality to these common libraries and utilities, in the hope that it will be beneficial to other teams. However, the problem it creates is that additional unused code is now being imported into the other micro apps (while tree shaking will solve this problem, in most cases, mainly due to poor coding practices, tree shaking doesn't work well, and we end up with unnecessary code imported into the apps). Another problem with these shared libraries is that there is a much higher risk of introducing breaking changes, with changes made for one micro app now breaking the other micro apps. By going overboard with code reusability, we end up with what's commonly called a "distributed monolith," which is essentially the worst of both worlds.

It's okay to have some shared libraries or, if using TypeScript, a shared types/interfaces file, but we must avoid creating large common libraries.

In the microservice/microfrontend world, DRY essentially refers to automating tasks so that you don't have to manually repeat the steps for each microservice or micro app. These could be things such as automating quality gates, or performance and security checks as part of the developers' pipelines.

Technology agnostic

Another principle of a microfrontend architecture is that it should be technology agnostic, meaning that each of the micro apps could "in theory" be built using different frameworks/languages. However, just because it's possible doesn't mean teams should go all out and use either Vue, Angular, or React to build out different micro apps.

There are multiple reasons why this should be avoided:

- Multiple libraries/frameworks mean an additional payload being sent down the wire to users' devices

- It makes it difficult to rotate team members, and moving from one team to another means having to get comfortable with a new framework/library

The primary reason for this principle is to allow for incremental upgrades, either from an older version to a newer version of the same library or to explore the benefits of a new framework.

Granular Scaling

When planning out a deployment strategy for your microfrontend, you must ensure that it supports granular scaling. By granular scaling, what we mean is that if a certain set of pages is getting a lot of traffic, either due to a marketing campaign or something similar, then only the servers serving those pages should scale, while the rest of the pods serving other parts of your microfrontend can remain at their regular levels. This ensures optimal cloud and hosting costs.

Culture of Automation and DevOps

A strong culture of automation and DevOps is critical for the long-term success of a microfrontend architecture.

As you can imagine with microfrontends, since we break up a single app into smaller apps, all the activities associated with tasks such as compiling the app and running quality, performance, and security checks will now need to be done multiple times for each of the apps. If we don't have automation processes for all of the aforementioned items, then the overall development and release of these apps will take a lot longer than what it would have been with a monolith.

Hence, it is important to invest time and effort into building these automation processes, most of which are generally done as part of the DevOps pipelines.

Teams can also invest in tooling and building code generators and micro app templates that can help speed up the creation of newer micro apps. They can also run linters, security, and other quality checks automatically as part of the DevOps pipelines.

With this, we come to the end of this section, where we saw some of the important principles that teams must keep in mind when designing a microfrontend architecture.

We saw how principles such as domain-driven teams, independent deployments, and granular scaling allow teams to move consistently and quickly. We saw how teams should avoid falling into the trap of a distributed monolith and build a pattern that uses build-time integrations, and finally, we saw how keeping the architecture technology agnostic and focusing on automation helps an architecture to easily evolve and become future-proof.

In the next section, we will look at some of the important components of the microfrontend architecture.

The key Components of a Microfrontend Architecture

After spending time going through the principles of a microfrontend, now let's look at some of the key components of a microfrontend architecture.

In this section, we will look at the essential components any microfrontend architecture needs to have, and we will look at some of the nuances associated with them.

After completing this section, you will be aware of the four basic components that make up any microfrontend architecture.

Routing Engine

As we saw in the previous chapter, depending on the type of microfrontend pattern you aim to build, the routing engine for your app will be partially or fully decoupled from your apps.

There are multiple approaches we can take. We can use NGINX as a reverse proxy and have a list of all the primary routes that map to the respective apps in the multi-SPA pattern. If the apps are deployed in a Kubernetes cluster, we can make use of Ingress routes to map the primary routes to the respective apps. We will go into more detail about this in *Chapter 8, Deploying Microfrontends to Kubernetes*, where we will look at deploying these microfrontends in the cloud.

A global state and a Communication Channel

In addition to routing, the next important thing to design well in your microfrontend architecture is the communication channel between the different apps and also the notion of a global state, which can be shared between the different apps.

With a monolith SPA, the most common practice is to use a single global store such as Redux or MobX, where everything is written into that store and read from it. With microfrontends, the recommendation is to avoid such global client-side stores and instead let each micro app get its data from the backend API, as that is the real source of truth.

However, there would be a genuine need for client-side state management to avoid making unnecessary calls to the backend, to fetch things such as `user_id` or a cart count. For things such as these, we can look to use a really thin global store in the app shell or maybe even look toward `localStorage` or `IndexedDB` to store the values that are needed to make API calls.

With a micro app microfrontend pattern, it also becomes important to establish a common communication channel that the different apps use to communicate with each other. A classic use case would be when clicking on the **Add to Cart** button on a product page, the mini cart present in the header is automatically incremented. In such cases, an event-driven communication channel works best.

Source code Version Control

Another important item that teams need to agree on is how they plan to organize their Git repositories. Two schools of thought prevail here – organizing your apps in a polyrepo or a monorepo. Let's look at their nuances.

Polyrepos

Polyrepos are where you have each of your multi-SPAs or micro apps managed in its own independent Git repository. These are easiest to start with and give complete team independence. From a DevOps standpoint too, they are a lot easier to manage. However, this approach has a few drawbacks. There is a higher risk of teams becoming siloed and reduced inter-team collaboration. Another drawback is duplication and higher maintenance costs for tooling, such as DevOps pipelines and automation scripts, which need to be duplicated and updated in each of the repos.

Monorepos

In a monorepo structure, all your multi-SPAs or micro apps are co-located in a single Git repo, with each app located within its own individual folder.

Monorepos are starting to become a de facto approach for many frontend teams to manage their code repositories. The main advantage of monorepos is increased team collaboration, as everybody is able to see every other team's code and provide valuable feedback. Tooling and automation scripts can be centralized, whereby optimizations done by one team are immediately available for other teams to follow. Some of the drawbacks of monorepos include DevOps setups being a bit complicated. Teams also need to set up fine-grained folder-level permissions to prevent teams from overwriting each other's code. In the grand scheme of things, monorepos provide more advantages and, hence, are a preferred approach to managing the source code for your microfrontends.

A Component Library

When building microfrontends, it is critical to ensure a consistent look and feel as a user navigates through the different apps. The way we achieve that is by ensuring all apps make use of a common design system and component library. It is also recommended that all teams use a common theming and styling engine to ensure that all the components look and behave the same, irrespective of which app they are served in.

A common pattern is to publish a component library as an NPM module and set up all the other apps to import and use it. Each time a new version of the component library is published, teams will need to update their respective apps to the latest version.

An emerging trend, thanks to monorepos, is to build directly from source. What this means is that a component library is stored within the `libs` section of the monorepo and the components are directly linked from the library path. The main advantage of this method is that every time teams build their app, they automatically receive the latest version of the component library.

In this section, we learned about the key components of a microfrontend architecture, namely a routing engine, a global state, and a communication channel. We also saw the distinctions between a polyrepo and monorepo and saw why frontend teams prefer to use monorepos. Finally, we also learned about the component library and different ways teams consume components from a common library.

Summary

With that, we come to the end of our second chapter. We started the chapter by looking at the key principles we need to keep in mind. We saw why it is important to break teams down based on domain models, and why it is critical for teams to be able to independently deploy their own apps. We learned about the misconceptions associated with code reuse and how it can lead to a distributed monolith trap. We also saw the importance of DevOps and an automation culture. Finally, we learned about the four key components of a microfrontend. Everything that we learned in this chapter we will put into practice in the coming chapters, as we go about building our very own microfrontend application.

In the next chapter, we will dive deeper into monorepos versus polyrepos and learn how it's more about team culture than technology. We will also start off by setting up our code repository as a monorepo to set up the foundation for future work.

3

Monorepos versus Polyrepos for Microfrontends

Since the time engineers at Google and Facebook mentioned they have a single monorepo in their organization, the developer community – especially the frontend community – has been actively participating in debates and discussions on monorepos versus polyrepos.

We are seeing more and more teams leaning more toward monorepos for maintaining their frontend code. However, which should you choose between a polyrepo and a monorepo based on what the community thinks?

As we will learn in this chapter, the decision to go with a monorepo or a polyrepo is far deeper than just fancy technology or hype. We will see that, in fact, it is more to do with teams, and the culture we would like to establish within teams.

In this chapter, we start by understanding what polyrepos and monorepos are. We will see how each of them impacts how teams work and collaborate, then we will see why monorepos are more suited for microfrontends. Finally, we will set up our monorepo base application with the necessary permissions to work in teams.

In this chapter, we will cover the following topics:

- Repo types and their nuances
- Why Monorepos for Microfrontends?
- Setting up our Monorepo with Team permissions

By the end of this chapter, you will have a deep understanding of the differences between and implications of choosing a polyrepo versus a monorepo.

We will also have our monorepo set up and ready for vertically sliced domain-driven teams to get started with it.

Technical requirements

As we go through the code examples in this chapter, we will need the following:

- A PC, Mac, or Linux desktop or laptop with at least 8 GB of RAM (16 GB preferred)
- An Intel chipset i5+, AMD, or Mac M1 + chipset
- At least 256 GB of free hard disk storage

You will also need the following software installed on your computer:

- Node.js version 16+ (use nvm to manage different versions of Node.js if you have to)
- Terminal: iTerm2 with Oh My Zsh (you will thank me later)
- IDE: We strongly recommend VS Code as we will be making use of some of the plugins that come with VS Code for an improved developer experience
- npm, yarn, or pnpm – we recommend pnpm because it's fast and storage-efficient
- Browser: Chrome, Microsoft Edge, Brave, or Firefox (I use Firefox)

The code files for this chapter can be found here: `https://github.com/PacktPublishing/Building-Micro-Frontends-with-React`

We also assume you have a basic working knowledge of Git, such as branching, committing code, and raising pull requests.

Repo types and their nuances

In this section, we will learn exactly what a polyrepo and a monorepo are.

As most of you will already know by now, repo is short for repository and refers to storage for all the files for your project. It also keeps track of all the changes to those files. This means, at any time, we can easily go and see what lines of code were changed, by whom, and when. In most cases, we use Git for version control. Some teams may use other systems, such as Mercurial or some other distributed version control system.

There are two strategies that teams most commonly use for managing repos. They are commonly known as monorepos and polyrepos. There are other patterns, such as Git submodules or Git subtrees, but these are beyond the scope of this chapter. We will focus on monorepos and polyrepos.

Monorepos

As the name suggests, mono means single and so the source code is managed in a single Git repo. This means that all team members work on a common single repository, and in most cases, the monorepo will consist of multiple applications. The following figure shows a monorepo setup:

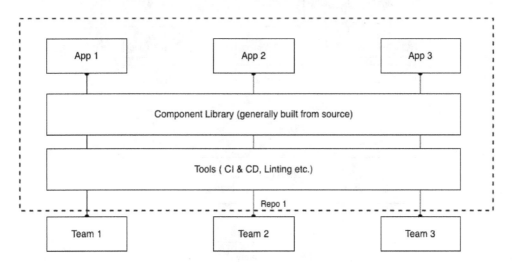

Figure 3.1 – Monorepo setup

As you can see in the preceding diagram, we have a single repo that consists of multiple apps within it. All the apps use a shared set of tools for CI and CD and linting and a shared component library that is usually built from source each time an application is built. You will also notice that all the teams have access to all the items within the repo.

Polyrepos

Polyrepos are where each app has its own repository. Teams generally work on multiple repos, switching repos as they work on different apps.

Most teams prefer going down the polyrepos route as they are a lot easier to manage and each team can define its own branching strategies and repo permissions. The following figure shows a polyrepo setup:

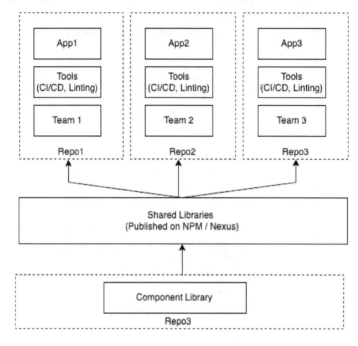

Figure 3.2 – Polyrepo setup

In the polyrepo setup, you will notice multiple repositories, denoted by the dotted-line boxes. Each app generally has its own repository, and we have another repository for shared components and libraries. The shared components need to first be published to an artifact repository such as npm or Nexus before they can be consumed by the other repositories. We also notice that each repository has its own team, and generally, teams don't have access to another team's repository (unless you are an admin or a senior developer that looks after multiple applications).

Differences between Polyrepos and Monorepos

As mentioned earlier, choosing between a polyrepo or a monorepo is not just about how code is organized but goes a lot deeper and has a huge impact on how teams collaborate, the culture within the teams, how your build tools are set up, and so on.

In this section, let us go a bit deeper into understanding the nuances of monorepos and polyrepos.

Team Collaboration

With polyrepos, teams create their own individual repositories and decide and define their own rules and guidelines on how the code is maintained. Obviously, this is the easiest and fastest way to get started, and teams become productive quite quickly. However, this pattern also has a few drawbacks. With polyrepos, teams tend to become more siloed as each team is just focused on its repo and doesn't really have a lot of visibility of what other teams are doing.

Another drawback with polyrepos is the effort required to set up and maintain all the build pipelines, and precommit hooks and so on are duplicated for each repo.

With monorepos, teams are forced to collaborate as they need to agree on a common way for how the code will be maintained. In a monorepo setup, since everybody is able to see everybody else's code, the chances of them working in silos are greatly reduced. Teams are naturally encouraged to collaborate by providing feedback on code and it also provides an opportunity for teams to replicate good code patterns that other teams may have implemented.

Build tools and Quality gates

With polyrepos, each team needs to implement its own build systems and quality gates such as precommit hooks. This results in duplication of effort and leads to higher maintenance costs. It also boils down to the engineering maturity of each team. Teams with strong leads will obviously have optimized build tools, while junior teams will struggle with not-very-optimized build tools and quality gates and will need intervention from other teams.

With a monorepo, all the build tools and quality gates can be centrally managed, reducing duplication of effort. In most cases, this is usually set up by the experts within one of the teams. This allows leveraging strengths and skillsets across all teams and teams immediately benefit from the knowledge within the wider organization.

Code Ownership

In a polyrepo, the permissions are set at a repo level in terms of who has permission to view the code in a repo or make changes to it.

In a monorepo, all team members have access to view and edit all the files in the code. The permissions and control in a monorepo are maintained via a **CODEOWNERS** file, which allows us to set granular permissions at a folder level within the monorepo. We will be learning more about the CODEOWNERS file later in this chapter.

The mental model with monorepos is everybody in the team can make changes to a file and raise a request to merge their changes; however, only the rightful owner defined in the CODEOWNERS file has the permission to accept or reject the changes being made by a team member.

Flexibility

As may be obvious by now, polyrepos provide the highest level of flexibility, in terms of how the code within each team is managed.

In a monorepo, this flexibility is intentionally restricted to ensure that all team members benefit from the best coding practices and tooling setup that the team can offer.

Refactoring Code

With polyrepos, refactoring code across multiple repos can be time-consuming as one will need to check out all the different repos and individually raise a merge request or pull request for each repository.

With monorepos, doing such large-scale refactoring is quite easy by making atomic commits where a single merge request can contain the necessary changes for all the apps.

Ownership of Upgrades

When the time comes to upgrade libraries or tooling is when the most interesting differences between a polyrepo and monorepo setup come into play.

In a polyrepo, the onus of upgrading shared libraries or tools lies with each of the teams, and the teams can choose to defer an upgrade if they have other priorities. This can be both good and bad. While it allows teams to upgrade at their own pace, there is always a risk that some teams may get far behind in upgrading their libraries. This becomes a serious issue if an outdated library has a serious security vulnerability, and teams have ignored upgrading it. Since each team is responsible for upgrading libraries, they are also responsible for fixing breaking changes, and this is often the primary trigger to defer upgrades.

With monorepos, if a shared library or tool is being upgraded, it is easy to make atomic commits across all the apps within the monorepo, which means all teams directly get the benefit of the latest versions. What's interesting with monorepos (*that have the right build tools and quality gates in place*) is the ownership of fixing any breaking changes lies with the library owner or the person doing the upgrade, as the build pipelines will not allow you to merge the code unless it passes all the build steps and quality gates.

Code base Size

With polyrepos, your code base gradually increases over time; however, with a monorepo, you are dealing with a large code base right from day one, and the monorepo tends to grow exponentially as the application grows.

A large code base has a negative impact on productivity. Not only does checking out code take time but also, all the other activities, such as running build steps or running unit tests, take longer both on the local developer PC and also on the CI and CD pipelines.

Unless one makes use of features such as caching and building and testing only what has changed, monorepos can become very slow to work with.

As we reach the end of this section, we have learned about the differences between polyrepos and monorepos and have gone into the details of how they differ when it comes to things such as code refactoring and ownership, tooling team culture, collaboration, and so on.

In the next section, we will see which of the two is more suited for building microfrontends.

Choosing Monorepos for Microfrontends

After going through the pros and cons of polyrepos and monorepos, which one would you choose to use for your project? Well, you can choose either one of them and build microfrontends. Like all things in programming, there are trade-offs for every decision you make, and you need to be clear about what trade-offs you are comfortable with.

For the rest of this book, we will choose to go with the monorepo setup for the following reasons:

- With monorepos, team members are naturally encouraged to collaborate by learning and reviewing each other's code.

- It allows all teams to easily use a shared library of components. This ensures that each micro-app built as part of the overall app has the same look and feel and the overall user experience is consistent as the user interacts with the different micro-apps.

- It also makes it easy for central platform teams, such as the DevOps team or admin team, to easily refactor code across all the micro-apps.

- Some of the drawbacks of monorepos, such as the slower execution of quality gates on pipelines, can be overcome by making use of caching techniques, many of which are the default with most monorepo tools.

- As your overall app grows, and new features get added, new micro-apps will keep getting added to your app. Now, if you have a polyrepo setup with each micro-app in its own repo, it will become quite difficult to manage the large number of repos.

- In a microfrontend setup, most of the time, you would work on your individual micro-app; however, at times you would need to run all the micro-apps together to test out your app locally. This would be quite difficult to achieve if your micro apps are set up in a polyrepo.

In the following section, we will have a look at some of the popular open source monorepo tools out there, which will help you decide which would be the most suitable for you.

Popular Monorepo tools

This section covers some of the most popular open source monorepo tools that you can choose when building your microfrontends.

Lerna

Lerna was probably the first and most widely used monorepo tool. It follows what is called the packages-based monorepo style. What this basically means is each app sits under the `packages` folder and has its own `package.json` file, so every app has its own set of dependencies and there is nothing common between these apps.

Lerna was recently adopted by the nrwl team who originally built the Nx monorepo.

Nx

Nx was the next monorepo to become very popular and is probably the most mature and feature-rich of all the monorepo tools out there. Nx started off as an integrated monorepo. What that means is, in Nx, there is a single `package.json` file on the root and all apps use the same version of the packages. Nx has now evolved to also support the package-based style of monorepos.

It comes with advanced local and distributed caching solutions and is ideal for managing large monorepo code bases.

Turborepo

Turborepo is the newest entrant in the monorepos war. It follows a package-based style and is very similar to how Lerna works. The main advantage of Turborepo is it supports a local and distributed caching system and is tightly integrated with Vercel's product suite, including Next.js and Vercel cloud hosting.

As we come to the end of this section, we have learned about the pros and cons of polyrepos versus monorepos. We saw some of the reasons why we choose to use monorepos for microfrontends and we also learned about some of the popular monorepo tools that teams use. In the next section, we will get our hands dirty setting up our monorepo.

Setting up our Monorepo

In this section, we are going to set up our monorepo, which will act as a base for our microfrontend apps. We will learn how to set up the right permissions and the necessary quality gates. Along the way, we will also learn about a couple of productivity tricks and plugins that improve the developer experience.

For this example and the rest of the chapters, we will use Nx as the monorepo to build our microfrontends as it allows you to build both a package-setup-style and an integrated-style monorepo. You can equally choose either Lerna or Turborepo to build your microfrontends.

Follow along with the step-by-step guide to set up an Nx monorepo:

1. Open up the terminal, `cd` into the folder where you generally keep your projects, and run the following command:

    ```
    pnpx create-nx-workspace@14
    ```

    The preceding command will install a couple of libraries and will then prompt you for the name of the workspace (e.g., `org name`). This will be the name of the folder within which your monorepo will be set up. We will call it **my-mfe** for the sake of consistency.

2. Next, it will prompt you to select what kind of apps you would like to create. We will choose **react**:

```
✓ Workspace name (e.g., org name)    · my-mfe
? What to create in the new workspace …
apps                [an empty workspace with no plugins with a layout that works best for building apps]
npm                 [an empty workspace with no plugins set up to publish npm packages (similar to yarn workspaces)]
ts                  [an empty workspace with the JS/TS plugin preinstalled]
react               [a workspace with a single React application]
angular             [a workspace with a single Angular application]
next.js             [a workspace with a single Next.js application]
nest                [a workspace with a single Nest application]
express             [a workspace with a single Express application]
web components      [a workspace with a single app built using web components]
react-native        [a workspace with a single React Native application]
react-express       [a workspace with a full stack application (React + Express)]
angular-nest        [a workspace with a full stack application (Angular + Nest)]
```

Figure 3.3 – Select a workspace with a single React application

3. When prompted for the application name, enter `catalog`, as this will be the catalog app within our microfrontend.

4. When prompted to select the stylesheet format, you can select the default, **CSS**, or any other format you prefer.

5. Next, it will prompt you to enable distributed caching. For this exercise, we will say **No**.

> **Important note**
> You can find complete details of setting up NX here: `https://nx.dev/getting-started/intro`

It will then go on to install all the dependencies and, once successfully completed, you should have a folder structure similar to *Figure 3.4*:

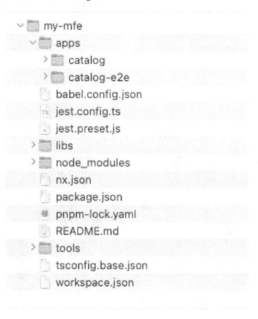

Figure 3.4 – Folder structure for our monorepo

You will notice it will have created a monorepo called my-mfe and an app called catalog within the apps folder.

6. Go ahead and open up this folder in Visual Studio Code and, once you do that, you will get a prompt to install recommended plugins. Go ahead and install the recommended plugins.

Once all the plugins have been installed, you will notice a new icon on the VS Code pane, as highlighted here:

Figure 3.5 – Nx Console installed on VS Code

Nx Console is one of the coolest features of using Nx and we will be extensively using it for the rest of this book.

For those curious about how this popup to install recommended plugins came up, the answer lies in the my-mfe/.vscode/extensions.json file.

This is a VS Code feature and you can read about it here: https://code.visualstudio.com/docs/editor/extension-marketplace#_workspace-recommended-extensions

You can use this file to add your own list of recommended plugins that you would like your team members to use.

This is an easy way for teams to standardize plugins and help junior developers get productive faster without them having to learn things the hard way.

You will also notice that Nx has also created a few other files, such as `eslintrc.json`, `.prettierrc`, `.editorconfig`, and so on. All of these files help lay a good foundation for writing good code and ensuring consistency in how that code is written with regard to things such as indentation, the use of single versus double quotes, and so on.

Running the app locally

To run the app locally, we could always run the terminal commands, but for a better developer experience, we will use the newly auto-added Nx Console extension we talked about earlier.

Click on the Nx Console icon and then, under **GENERATE & RUN TARGET**, select **serve** and then, from the dropdown at the top, select the **catalog** app, then select **Execute:nx run catalog:serve**

Figure 3.6 – Serving the catalog app using Nx Console

You will notice that it actually runs `pnpm exec nx serve catalog` in the terminal and, after a few seconds, you will have the catalog app running at `http://localhost:4200`.

Open the link in the browser and get a feel for the newly created catalog app:

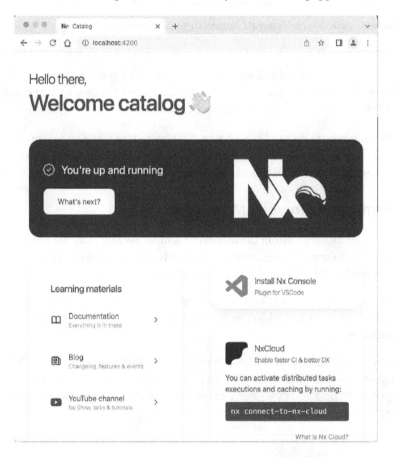

Figure 3.7 – Catalog app running on port 4200

Creating a new app with Nx Console

Next, let us create another new app. Follow these steps:

1. Go to Nx Console and from **GENERATE & RUN TARGET**, select the **generate** command. Then, from the dropdown, select **Create React application**. On the following screen, where it asks for the name of the application, enter `checkout`.

2. As you scroll down the form for the section that says **e2eTestRunner**, select none. This will ensure the `checkout-e2e` folder is not created.

> **Important note**
>
> Notice as you fill in the form fields that Nx is actually doing a dry run on the terminal to show what the output would look like.

3. Go ahead and click on the **Generate** button to generate our checkout app.

    Once done, you will notice the checkout app created within the `apps` folder of the monorepo.

4. Using Nx Console, go ahead and serve the checkout app. On the nx serve checkout screen, scroll down a bit and select **port** and type 4201 then select **Execute:nx run checkout:serve –port=4201** to run the checkout app on port 4201.

We can follow the same steps to create additional apps. Nx comes with a whole set of core and community plugins, which allows you to create apps in different frameworks, such as Angular, Next.js, Vue, and so on. You can view the full list of plugins available here: `https://nx.dev/community`.

Setting permissions in your Monorepo

Now that we have multiple apps within the monorepo and we assume there are independent teams working on each of these apps, the next thing that arises is how we ensure the right permissions to ensure that teams don't accidentally make changes to another team's code.

As we saw earlier, the general thought process with monorepos is that everybody with access to the repo has access to all the apps and folders within the monorepo but they can't merge code changes in apps they don't own.

In monorepos, the permissions are set at a folder level and by making use of the `CODEOWNERS` file. You can read in detail about `CODEOWNERS` here: `https://docs.github.com/en/repositories/managing-your-repositorys-settings-and-features/customizing-your-repository/about-code-owners`.

> **Important note**
>
> The `CODEOWNERS` file works with GitHub and GitLab. If you are using Azure DevOps, this feature is implemented via the required approvers feature: `https://learn.microsoft.com/en-gb/azure/devops/repos/git/branch-policies?view=azure-devops&tabs=browser#automatically-include-code-reviewers`.

In short, the `CODEOWNERS` file allows us to ensure that an individual or a team is explicitly involved in the code review and approval of changes to files they own.

We can assign files in two ways: all files within a given folder or all files of a certain type.

Let's see this in action.

In the root of our monorepo, let's create a file called CODEOWNERS:

```
/apps/catalog @my-org/catalog-team
/apps/checkout @my-org/checkout-team
```

What this means is if any of the pull requests contain modifications to files within the apps/catalog folder, it will automatically add people from the catalog team as reviewers to the pull request, and without approval from that team, the pull request cannot be merged.

The same holds true for pull requests with modifications to files in the checkout folder. In this case, it will require an explicit approval from members of the checkout team.

We can also assign an individual in the CODEOWNERS file. Let's say we want to ensure that any changes to files in the tools folder need approval from GitHub user @msadmin. Let's also assume we have a CSS expert on our team and would like that person to review all CSS changes in the entire repo. We can add the following two rules to enable this:

```
/tools @msadmin
*.css @cssexpert
```

This way, we can ensure a fine-grained approval process for pull requests, ensuring that the right stakeholders are involved in the approval of all changes being made to the files they are responsible for. As you can see, this also allows you to set rules so that an individual's expertise on a certain subject can be leveraged for the overall benefit of the whole team.

The following are a few points to keep in mind as you create entries in the CODEOWNERS file:

- File paths in the file are case-sensitive
- The priority of the rules is from the bottom to the top of the CODEOWNERS file; for example, if there are multiple matching rules, the bottom-most row gets the highest priority
- If a row has a syntax error, it will be skipped, and GitHub will simply move on to the next row

To test this, push the code to GitHub with the entries in the CODEOWNERS file, make changes to the file, and raise a pull request to see the CODEOWNERS file come into action.

Having come to the end of this section, we have learned how to initialize a monorepo using Nx, how to create apps within our monorepo, and how to run them individually using NX Console. We also had a quick look at some of the tooling advantages we get with Nx, which offers a really good developer experience for beginners and also ensures a strong foundation for your application by automatically setting up some of the quality gates for your repo. Finally, we looked at the various way we can set up permissions on our repo to allow open collaboration and also leverage individual team members' strengths for the benefit of the whole team.

Summary

With this, we come to the end of this chapter, where we unpacked quite a bit. We saw how teams today are choosing between a monorepo and a polyrepo approach to version control their code bases.

We then went into the details of how the choice of a polyrepo or a monorepo impacts how your teams operate, how easy or difficult it is to refactor code, and who owns the responsibility of fixing breaking changes in the repo.

We then saw why choosing monorepos for microfrontends has more benefits, such as the ease of managing all the micro-apps within a single repo, especially when it comes to running multiple apps locally and managing the tooling centrally for all the apps within the monorepo.

Finally, we went about setting up our monorepo, where we saw the benefits of using a tool such as Nx, which provides us with prebaked quality gates such as ESLint and Prettier to ensure consistency and code quality. We also saw how to use Nx Console to easily create new micro-apps and run existing micro-apps. We then saw how to set up the CODEOWNERS file to ensure granular control over who can approve code changes for a given micro-app.

In the next chapter, we will take our current setup and go about creating a full-fledged multi-SPA pattern microfrontend.

Part 2:
Architecting Microfrontends

This part explores various architectural patterns for implementing microfrontends, including Multi-SPA, micro-apps, Module Federation, and server-side rendering approaches. It provides concrete examples that cover topics around routing and state management for microfrontends.

This part has the following chapters:

- *Chapter 4, Implementing the Multi-SPA Pattern*
- *Chapter 5, Implementing the Micro-Apps Pattern*
- *Chapter 6, Server - Rendered Microfrontends*

4

Implementing the Multi-SPA Pattern for Microfrontends

Imagine you are an architect tasked with building the frontend for a large government ePortal that has and provides numerous online services for individuals and businesses. These services include registering for health benefits, submitting accounts for income tax, registering a small business, and paying vehicle road tax, in addition to publishing a whole bunch of informational content.

Or, scenario two, imagine you have been tasked to build a banking portal that provides multiple online services, from managing saving accounts to buying insurance, to investment opportunities, loans, mortgages, credit cards, and so on.

How would you go about planning not just your architecture but also the team that will be responsible for building it? Naturally, the first level of thinking would be to break down the large portal into multiple smaller modules or mini-apps and have each team focus on one of these mini-apps.

This would be the right approach, and this is also what we refer to as the multi-SPA pattern for building microfrontends.

In this chapter, we will go about building our multi-SPA pattern microfrontend, where we will look at the following:

- The high-level architecture of the multi-SPA microfrontend
- Establishing routing between multi-SPAs
- Using a shared component library
- Setting up a persistent state to share state between mini apps

Technical requirements

As we go through the code examples in this chapter, we will need the following:

- A PC, Mac, or Linux desktop or laptop with at least 8 GB of RAM (16 GB preferred)
- An Intel chipset i5+ or Mac M1+ chipset
- At least 256 GB of free hard disk storage

You will also need the following software installed on your computer:

- Node.js version 16+ (use nvm to manage different versions of Node.js if you have to)
- Terminal: iTerm2 with OhMyZsh (you will thank me later)
- IDE: We strongly recommend VS Code as we will be making use of some of the plugins that come with VS Code for an improved developer experience
- npm, yarn, or pnpm – we recommend pnpm because it's fast and storage-efficient
- Browser: Chrome, Microsoft Edge, or Firefox (I use Firefox)
- A basic understanding of Nx.dev monorepos and a basic understanding of using the NX Console plugin in VS Code
- Working knowledge of React

The code files for this chapter can be found here: `https://github.com/PacktPublishing/Building-Micro-Frontends-with-React`

We also assume you have a basic working knowledge of Git, such as branching, committing code, and raising pull requests.

Understanding the multi-SPA architecture

The multi-SPA architecture pattern is one of the most common patterns for building large-scale applications. As the name suggests, in this pattern, we have a collection of SPAs that together form a large application. In this pattern, each SPA behaves as its own independent feature or module that can be directly accessed via a URL namespaced and mapped to the app. These SPAs also share a very thin layer of shared components and global state to ensure coherency and consistency between the apps.

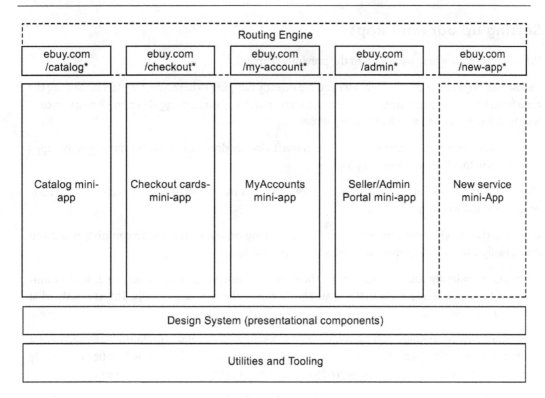

Figure 4.1 – The multi-SPA architecture

As you can see in *Figure 4.1*, we have four SPAs: a catalog, which will hold pages such as product listings, product details, search, and so on; a checkout SPA containing pages such as shopping cart, payments, and so on; the MyAccounts SPA; and the Seller/Admin SPA. You will also notice that this pattern allows us to easily add additional SPAs as the application grows.

Each of these SPAs is mapped to a unique primary URL, such that users clicking the /catalog URL will be redirected to the catalog app, while users clicking the /checkout URL will go to the checkout app.

Building our Multi-SPA Microfrontend

Building a multi-SPA microfrontend essentially consists of three broad areas: breaking down the app into logical mini-apps, then we need to set up the routing between these mini-apps, and finally, we set up a global state the different mini-apps can read and write data to. Let us look at each of them in the subsequent sections.

Setting up our mini-apps

We will start with where we left off in the previous chapter.

In case you skipped the previous chapter and are directly jumping in here, you can start by cloning the repo from `https://github.com/PacktPublishing/Building-Micro-Frontends-with-React/tree/main/ch3/my-mfe`.

Let us quickly run `pnpm install` (if you haven't already done so) and serve the respective apps to make sure that they are running properly.

Since we are going to build an e-commerce application, let us call our app *eBuy*. Feel free to rename your app folder to `ebuy`.

During active development, we would ideally be working on our own respective mini-app and you could easily use the NX Console to serve your respective app.

However, periodically you may want to test the entire end-to-end app flow across the different mini-apps and for that, it is important that you are able to run all the mini-apps locally. This is exactly what we are going to do next.

We first need to ensure that each mini-app runs on its own unique port. To do this, we need to first locate the `project.json` file located in the `apps/catalog` folder. You will notice it basically contains all the commands and configuration needed to run the various tasks on your app.

We navigate to the `"serve":` section and under `"options"`, add the line `"port": 4200`:

```
"serve": {
  "executor": "@nrwl/web:dev-server",
  "defaultConfiguration": "development",
  "options": {
    "buildTarget": "catalog:build",
    "hmr": true,
    "port": 4200
  },
  "configurations": {
    "development": {
      "buildTarget": "catalog:build:development",
    },
    "production": {
      "buildTarget": "catalog:build:production",
      "hmr": false
    }
```

```
    }
  },
```

We do the exact same thing in the `project.json` file located in the `apps/checkout` folder, but this time we will ensure this runs on `"port": 4201` like so:

```
"options": {
  "buildTarget": "checkout:build",
  "hmr": true,
  "port": 4201
},
```

This will ensure that, by default, the catalog will run on port `4200` while the checkout app runs on port `4201`.

Thanks to inheritance, we will be able to run the app in development and production mode from the same ports.

Next, we will create a script command that will allow us to run all the apps in parallel on their respective ports.

For this, we go into the `package.json` file located at the root of the project and add a script called `"serve:all": "nx run-many --target=serve"`:

```
"scripts": {
  "start": "nx serve",
  "build": "nx build",
  "test": "nx test",
  "serve:all": "nx run-many --target=serve"
},
```

Now, in your terminal, run the following command:

```
pnpm serve:all
```

You will see `nx` is starting up the webpack development server and is launching the two apps.

Verify it by visiting these two URLs in the browser:

- Catalog app: `http://localhost:4200`
- Checkout app: `http://localhost:4201`

With microfrontends, it is important that each SPA follows the same brand guidelines and look and feel. We ensure this by building a shared set of UI components that both apps make use of. In the next section, we will see how to create a shared component library.

Using a shared component library

As you are building a series of mini-apps as part of your overall bigger app, we want to ensure that all these mini-apps have a consistent design – things such as having a consistent header and footer and a consistent way for the various components to behave. What is equally important is, when we make a change to some of these core elements, we need to ensure that it can be updated across all the different apps without too much trouble. This is where the libs folder comes into play.

This would also be a good time to define an NPM scope so that all these shared components can be imported via their scope names.

To define an NPM scope, we open up the nx.json file located at the root of the monorepo. We are going to name our scope ebuy but in reality, it could be anything – the name of your team, the name you have for your component library, and so on.

Locate the npmScope property in the nx.json file and update it as follows:

```
"npmScope": "ebuy",
```

Let us use our trustworthy Nx Console to create a library. From Nx Console, select generate and then select @nrwl/react - library React Library.

Select the **Show all options** and provide/modify the following details and leave the rest as the default:

```
Library name    : ui
Generate a default component    : No
importPath : @ebuy/ui
```

We can leave the rest as the default and click the run button to generate the ui folder within libs.

In addition to creating the ui folder within libs, you will notice Nx has also added an entry into the paths object of tsconfig.base.json as follows:

```
"paths": {
  "@ebuy/ui": ["libs/ui/src/index.ts"]
}
```

It is this setting that will allow us to import our UI components via the scoped name instead of a long folder path.

Next, let's create a couple of UI components.

We will use the awesome Semantic-UI React component library to build out our UI components. You can also use any other component library, such as Chakra UI, MUI React-Bootstrap, and so on:

1. Let's install it on the root of the monorepo using the following command:

    ```
    pnpm install semantic-ui-react semantic-ui-css
    ```

2. Remember you can always use npm or `yarn` to install npm packages as follows:

    ```
    yarn add semantic-ui-react semantic-ui-css
    ```

    ```
    npm install semantic-ui-react semantic-ui-css
    ```

    Now let's create a couple of our common components in the `libs/ui` folder.

3. Let us use Nx Console and create a new component:

    Nx | Generate | Create a react component

4. Use the following information to create the component:

    - **Name**: header

    - **Project**: ui

    - **Flat**: Select the checkbox to ensure we have a flatter folder structure within

5. Hit the run button and verify the `header.tsx` file is created within the `libs/ui/src/lib` folder.

6. Open the `header.tsx` file and replace the contents of it with simple markup for our header component:

    ```
    import { Menu, Container, Icon, Label } from 'semantic-
    ui-react';

    export function Header() {
      return (
        <Menu fixed="top" inverted>
          <Container>
            <Menu.Item as="a" header>
              eBuy.com
            </Menu.Item>
            <MenuItems />
            <Menu.Item position="right">
              <Label>
    ```

```
                    <Icon name="shopping cart" />
                    00
                </Label>
            </Menu.Item>
        </Container>
    </Menu>
  );
}
const MenuItems = () => {
  return (
    <>
      {NAV_ITEMS.map((navItem, index) => (
        <Menu.Item key={index}>
          <a href={navItem.href ?? '#'}>{navItem.label}</
a>
        </Menu.Item>
      ))}
    </>
  );
};

interface NavItem {
  label: string;
  href?: string;
}

const NAV_ITEMS: Array<NavItem> = [
  {
    label: 'Catalog',
    href: '/',
  },
  {
    label: 'Checkout',
    href: '/checkout',
  },
];
export default Header;
```

This is simple React component code that will display the header with navigation for the catalog and checkout.

7. The next step is to export it out it from the `ui`. Locate the `/libs/ui/src/index.ts` file and add an entry as follows:

```
export * from './lib/header';
```

This will allow our header component to be importable via our shorter import path. Now let us import it into our catalog and checkout apps.

8. Open the `apps/catalog/src/spp/app.tsx` file and import the header component as follows:

```
import { Header } from '@ebuy/ui';
```

9. Let us clean up some of the boilerplate code. Remove the imports for `styles` and `NxWelcome` and add the `Header` component in the JSX. You can also delete the `nx-welcome.tsx` file in the `catalog` folder. Your final code should look like this:

```
import { Header } from '@ebuy/ui';
import { Container, Header as Text } from 'semantic-ui-
react';
import 'semantic-ui-css/semantic.min.css';
export function App() {
  return (
    <Container style={{ marginTop: '5rem' }}>
      <Header />
      <Text size="huge">Catalog App</Text>
    </Container>
  );
}


export default App;
```

In the preceding code, we import the semantic-ui's css file and include our `Header` component and text that displays the name of the app.

When running in the browser, the catalog app will look something like this:

Catalog App

Figure 4.2 – Catalog app with the common header menu bar

10. We will make the same changes to the `apps/checkout/src/app/app.tsx` file within the checkout app.

11. Let us test out our code. Run `pnpm serve:all` and refresh your browser on `http://localhost:4200` to see our latest changes.

Try clicking on the navigation links for the catalog or checkout and notice it doesn't do anything. That is because we haven't set up routing between our apps, which is exactly what we will be doing next.

Setting up Routing

As we discussed earlier, from time to time, we would like to test our end-to-end app functionality, and although we are able to run apps in parallel on different ports, there are some challenges with testing end-to-end functionality:

- We need to ensure a consistent navigation structure for our apps both on localhost and on production.

- Apps running on different ports are treated as apps on different domains and hence it will not be possible to share cookies, session states, and so on

To overcome these problems, we need to make the browser think the apps are running on the same port. We do this by setting up a reverse proxy. The way we will set up routing is each mini-app will have its own namespaced primary route, for example:

- `eBuy.com`: Home page app

- `eBuy.com/catalog`: Catalog app

- `eBuy.com/checkout`: Checkout app

The secondary routes are generally set up within the mini-apps themselves. For example, the product details page for, say, apples would be `eBuy.com/catalog/apples`.

Webpack development servers and Nx come with easy-to-use proxy support that we can take advantage of.

At the root of the catalog app, `/apps/catalog`, let us create a new file called `proxy.conf.json` with the following entries:

```
{
  "/catalog": {
    "target": "http://localhost:4200"
  },
  "/checkout": {
    "target": "http://localhost:4201"
  }
}
```

Next, we need to tell the catalog app to use this file for its proxy configuration.

We do this by adding the `proxyConfig` property to the development configuration under the serve object in the `apps/catalog/project.json` file as follows:

```
"options": {
    "buildTarget": "catalog:build",
    "hmr": true,
    "port": 4200,
    "proxyConfig": "apps/catalog/proxy.conf.json"
},
```

Let us quickly test it out. We will need to restart our development servers to pick up the latest proxy configurations.

Run the `serve:all` command and try clicking on the **Checkout** and **Catalog** navigation links... Erm... It didn't work and the same catalog app shows up when you click on the **Checkout** link... But wait – the title tag on the browser tab does show **Checkout**:

Figure 4.3 – Checkout app in the title but loading the catalog bundle

So, what's happening here? If you look at the development tools, the problem becomes quite obvious. What's happening here is the proxy has correctly redirected us to the checkout app and that's why we see the correct `index.html` file served via the checkout app, however, the script's `src` tags loading the `js` bundles point to the root and hence they are actually loading the `js` bundles from the catalog app.

Fixing this problem is relatively easy again thanks to Nx.

We simply need to define the `baseRef` for the checkout app. We do this by adding `"baseHref":` `"/checkout/"` to the `/apps/checkout/project.json` file.

This is what your development object under the parent serve object should look like:

```
"options": {
    "buildTarget": "checkout:build:development",
    "port": 4201,
    "baseHref": "/checkout/"
},
```

Restart the development servers and now you will be able to navigate between the two applications and have the right JS bundles load in. In the next section, we will work toward adding a product list response to simulate the mocked response from a product list API call.

Setting up a mocked product list

A common practice with all web development activities is to set up a mock server or a mocked set of API responses that the frontend apps can consume until the actual APIs are ready. Since our e-commerce app requires a list of products that will be required across all the other mini-apps, we create a shared library to hold our mocks.

So again, using our favorite, Nx Console, let us create another React library, let us call it `mocks`, and we will use the scope name `@ebuy/mocks`.

Within the `mocks` library at `libs/mocks/src/lib`, let us create our file called `product-list-mocks.tsx` with the following code:

```
interface productListItem {
   id: string;
   title: string;
   image: string;
   price: number;
}

export const PRODUCT_LIST_MOCKS: Array<productListItem> = [
```

```
  {
    id: '1',
    title: 'Apples',
    image: '/assets/apple.jpg',
    price: 1.99,
  },
  {
    id: '2',
    title: 'Oranges',
    image: '/assets/orange.jpg',
    price: 2.5,
  },
  {
    id: '3',
    title: 'Bananas',
    image: '/assets/banana.jpg',
    price: 0.7,
  },
];

export default PRODUCT_LIST_MOCKS;
```

Let us not forget to export it out from the /libs/mocks/src/index.ts file with the following line of code:

```
export * from './lib/product-list-mocks';
```

Also, don't forget to place the product images in the catalog app's src/assets folder. You can find the images here https://github.com/PacktPublishing/Building-Micro-Frontends-with-React-18/tree/main/ch4/ebuy/apps/catalog/src/assets.

We will now look to use this across our apps, wherever we need data from the product list.

Adding the product grid and checkout components

Now we have a decent-looking header and an app where we can navigate from one mini-app to the other. However, the rest of the app doesn't do much, so let's add a product list component to the catalog app and a shopping basket component to the checkout app.

We will start by creating the `ProductList` component within our `/apps/catalog/src/app` folder. We will name the file `product-list.tsx`. We will start by creating an empty shell component:

```
import { Card } from 'semantic-ui-react';
import ProductCard from './product-card';
import { PRODUCT_LIST_MOCKS } from '@ebuy/mocks';

export function ProductList() {
  return (
    <Card.Group>
      {PRODUCT_LIST_MOCKS.map((product) => (
        <ProductCard key={product.id} product={product} />
      ))}
    </Card.Group>
  );
}
export default ProductList;
```

We will get an error for the missing `ProductCard` component. Don't worry – we will create that component in the next step. Next, we need to create our `ProductCard` component. We will name the file `product-card.tsx`.

We start by defining the skeleton of our `ProductCard` component:

```
import { Button, Card, Image } from 'semantic-ui-react';

export function ProductCard(productData: any) {
  const { product } = productData;
  return (
    <Card>
      <Card.Content>
        <Image alt={product.title} src={product.image} />
        <Card.Header>{product.title}</Card.Header>
        <Card.Description>{product.description}</Card.
Description>
        <Card.Header>${product.price}</Card.Header>
      </Card.Content>
      <Card.Content extra>
        <div className="ui three buttons">
```

```
          <Button basic color="red">
            Remove
          </Button>
          <Button basic color="blue">
            {0}
          </Button>
          <Button basic color="green">
            Add
          </Button>
        </div>
      </Card.Content>
    </Card>
  );
}


export default ProductCard;
```

Next, let us import the `ProductList` `app.tsx` file of the catalog app located at `/apps/catalog/src/app/app.tsx`.

Your `app.tsx` code should now look like this:

```
import { Header } from '@ebuy/ui';
import { Container, Header as Text } from 'semantic-ui-react';
import 'semantic-ui-css/semantic.min.css';
import ProductList from './product-list';
export function App() {
  return (
    <Container style={{ marginTop: '5rem' }}>
      <Header />
      <Text size="huge">Catalog App</Text>
      <ProductList />
    </Container>
  );
}
export default App;
```

If your catalog app looks like the following screenshot, that means you are on the right path:

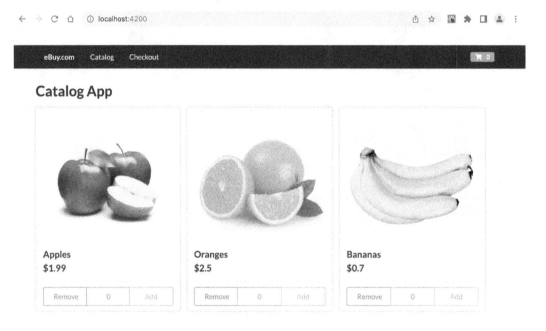

Figure 4.4 – Catalog app with a header and product list component

Next, we are going to create our shopping basket component. So, in our app.tsx checkout file located in the /apps/checkout/src/app folder, let us create a basic skeleton with the following code:

```
import { Header } from '@ebuy/ui';
import { Container, Header as Text } from 'semantic-ui-react';
import 'semantic-ui-css/semantic.min.css';
import ShoppingBasket from './basket';

import { PRODUCT_LIST_MOCKS } from '@ebuy/mocks';

export function App() {
  return (
    <Container style={{ marginTop: '5rem' }}>
      <Header />
      <Text size="huge">Checkout</Text>
      <ShoppingBasket basketList={PRODUCT_LIST_MOCKS} />
```

```
    </Container>
  );
}
export default App;
```

This code should start looking familiar now. As you can see, we have a `ShoppingBasket` component and, for the time being, we are passing `PRODUCT_LIST_MOCKS` to it for the purpose of mocking.

Next up is to create that `ShoppingBasket` component, which is throwing an error at the moment.

So let us create a `basket.tsx` file in the `/apps/checkout/src/` app folder:

```tsx
import { Table, Image, Container } from 'semantic-ui-react';

export function ShoppingBasket(basketListData: any) {
  const { basketList } = basketListData;
  return (
    <Container textAlign="center">
      <Table basic="very" rowed>
        <Table.Header>
          <Table.Row>
            <Table.HeaderCell>Items</Table.HeaderCell>
            <Table.HeaderCell>Amount</Table.HeaderCell>
            <Table.HeaderCell>Quantity</Table.HeaderCell>
            <Table.HeaderCell>Price</Table.HeaderCell>
          </Table.Row>
        </Table.Header>

        <Table.Body>
          {basketList.map((basketItem: any) => (
            <Table.Row key={basketItem.id}
>
              <Table.Cell>
                <Image src={basketItem.image} rounded
size="mini" />
              </Table.Cell>
              <Table.Cell> {basketItem.title}</Table.Cell>
              <Table.Cell>{basketItem.quantity || 0}</Table.
Cell>
```

```
            <Table.Cell>${basketItem.price * basketItem.
quantity}</Table.Cell>
            </Table.Row>
        ))}
      </Table.Body>
    </Table>
  </Container>
  );
}


export default ShoppingBasket;
```

This is all self-explanatory dummy markup content that at the moment doesn't do much. In the following sections, we are going to make this all work together.

Your running checkout app should now look like this:

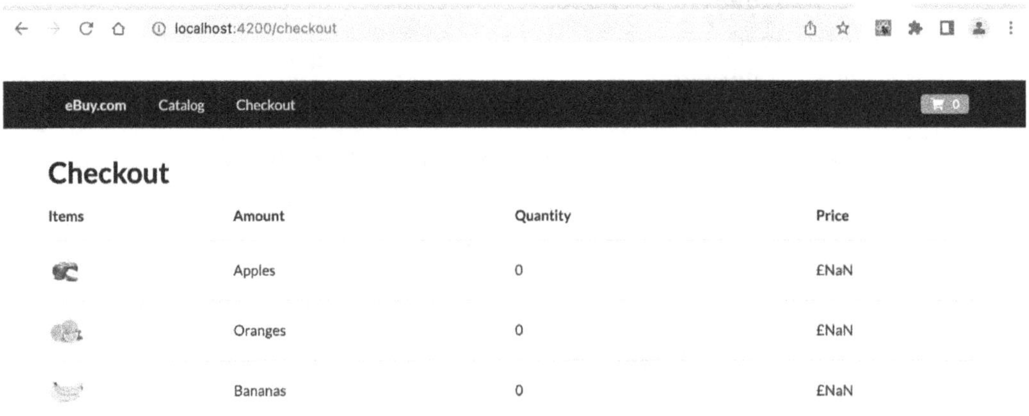

Figure 4.5 – Mocked up checkout app

With this, we have our two apps working well and displaying the right data, however, they are not "talking" to each other yet. The checkout app has no idea what items the user has added to the cart in the catalog app. In the next section, we will set up a global shared state that both the mini-apps can talk to and read from.

Before we proceed to the next section, let us quickly go through a checklist of things we've done so far:

- Ensured we have the catalog and checkout apps running on different ports
- Ensured we have the URL routing setup in the `proxy.conf.json` file
- We have both apps reading data from the mocked product list

Setting up a Global Shared State

Now that we are able to navigate between our two mini-apps, the next thing to tackle is setting up a shared state between these two different apps. Because these are two independent apps, the usual state management solutions such as the Context API, Redux, MobX, and so on will not work. This is because these libraries store the state as an object within the app and when you refresh the page or navigate to another app, this state is lost state. Hence, to overcome this problem, we resort to using some of the browser's native features, such as local storage, session storage, or Index-db.

For this example, we will be using session storage. We will set up a simple custom hook to persist state in `sessionStorage` and have both our mini-apps read and write to this state.

In any large-scale app, there will be a lot of similar custom hooks that teams can reuse. This is also a good opportunity for us to set up another library for these custom hooks.

It is important to remember that this global state should be used sparingly only when we need to share information between the different mini-apps. To manage the states within each micro app, we should use a regular state management tool such as the Context API or Redux, and so on.

Let us use Nx Console to create another library called `custom-hooks`:

```
Nx Console > generate > Create a React Library
```

Then, we'll fill use the following information in the form:

- **Name**: `custom-hooks`
- **Component**: `off` (Generate a default component)
- **importPath**: `@ebuy/custom-hooks`

Verify that the `custom-hooks` folder is created under `libs` and also make sure it has been added to the `tsconfig.base.json` file at the root of the monorepo, which should now look something like this:

```
"paths": {
  "@ebuy/custom-hooks": ["libs/custom-hooks/src/index.ts"],
  "@ebuy/mocks": ["libs/mocks/src/index.ts"],
  "@ebuy/ui": ["libs/ui/src/index.ts"],
```

```
      "@ebuy/utils": ["libs/utils/src/index.ts"]
  }
```

Let us now create our custom hook. Use the `generate` command to create a React component with the following information:

- **Name of the component**: `useSessionStorage`

- **Project**: `custom-hooks`

- **fileName**: `use-session-storage`

- **flat**: `Selected` (generate flat file structure)

In the newly created `use-session-storage.tsx` component file, let's replace the boilerplate code with the following:

```
import {
  Dispatch,
  SetStateAction,
  useCallback,
  useEffect,
  useState,
} from 'react';

import { useEventCallback, useEventListener } from 'usehooks-
ts';

declare global {
  interface WindowEventMap {
    'session-storage': CustomEvent;
  }
}
type SetValue<T> = Dispatch<SetStateAction<T>>;
export function useSessionStorage<T>(key: string, initialValue:
T): [T, SetValue<T>] {
  // Get from session storage then
  // parse stored json or return initialValue
  const readValue = useCallback((): T => {
    // Prevent build error "window is undefined" but keep
working
```

```
    if (typeof window === 'undefined') {
      return initialValue;
    }
    try {
      const item = window.sessionStorage.getItem(key);
      return item ? (parseJSON(item) as T) : initialValue;
    } catch (error) {
      console.warn(`Error reading sessionStorage key
"${key}":`, error);
      return initialValue;
    }
  }, [initialValue, key]);
  // State to store our value
  // Pass initial state function to useState so logic is only
executed once
  const [storedValue, setStoredValue] = useState<T>(readValue);
  // Return a wrapped version of useState's setter function
that ...
  // ... persists the new value to sessionStorage.
  const setValue: SetValue<T> = useEventCallback((value) => {
    // Prevent build error "window is undefined" but keeps
working
    if (typeof window === 'undefined') {
      console.warn(
        `Tried setting sessionStorage key "${key}" even though
environment is not a client`
      );
    }
    try {
      // Allow value to be a function so we have the same API
as useState
      const newValue = value instanceof Function ?
value(storedValue) : value;

      // Save to session storage
      window.sessionStorage.setItem(key, JSON.
stringify(newValue));
      // Save state
```

```
      setStoredValue(newValue);
      // We dispatch a custom event so every useSessionStorage
hook are notified
      window.dispatchEvent(new Event('session-storage'));
    } catch (error) {
      console.warn(`Error setting sessionStorage key
"${key}":`, error);
    }
  });
  useEffect(() => {
    setStoredValue(readValue());
    // eslint-disable-next-line react-hooks/exhaustive-deps
  }, []);
  const handleStorageChange = useCallback(
    (event: StorageEvent | CustomEvent) => {
      if ((event as StorageEvent)?.key && (event as
StorageEvent).key !== key) {
        return;
      }
      setStoredValue(readValue());
    },
    [key, readValue]
  );
  // this only works for other documents, not the current one
  useEventListener('storage', handleStorageChange);
  // this is a custom event, triggered in
writeValueTosessionStorage
  // See: useSessionStorage()
  useEventListener('session-storage', handleStorageChange);
  return [storedValue, setValue];
}
export default useSessionStorage;
// A wrapper for "JSON.parse()" to support "undefined" value
function parseJSON<T>(value: string | null): T | undefined {
  try {
    return value === 'undefined' ? undefined : JSON.parse(value
?? '');
```

```
  } catch {
    console.log('parsing error on', { value });
    return undefined;
  }
}
```

This custom hook code is part of the usehooks-ts library and is available here: https://usehooks-ts.com/react-hook/use-session-storage

Since this custom hook makes use of the usehook-ts library, we will install that npm module:

```
pnpn i usehook-ts
```

Next, we need to export it so that it can be imported via the scoped path. We do this in the /libs/custom-hooks/src/index.ts file by adding the following line:

```
export * from './lib/use-session-storage'
```

Next, we will use our newly created custom-hook in the product-card component such that every time the user adds products to or removes products from the shopping cart, it will store it as an array in sessionStorage.

In the /apps/catalog/src/app/productcard.tsx file, we will start by importing the useSessionStorage hook:

```
import { useSessionStorage } from '@ebuy/custom-hooks;
```

Then, within the product card component, we make use of the useSessionStorage hook and add the functions to add and remove items from the basket with the following code:

```
const [basket, setBasket]: any =
useSessionStorage('shoppingBasket', {});
  const addItem = (id: string) => {
    basket[id] = basket[id] ? basket[id] + 1 : 1;
    setBasket(basket);
  };

  const removeItem = (id: string) => {
    basket[id] = basket[id] <= 1 ? 0 : basket[id] - 1;
    setBasket(basket);
```

Next, we update the **Add** and **Remove** button on-click events as follows:

```
<div className="ui three buttons">
        <Button basic color="red" onClick={() =>
removeItem(product.id)}>
            Remove
        </Button>
        <Button basic color="blue">
            {basket[product.id] || 0}
        </Button>
        <Button basic color="green" onClick={() =>
addItem(product.id)}>
            Add
        </Button>
    </div>
```

Let's test this out by running the following command:

```
pnpm serve:all
```

Click on the **Add** and **Remove** buttons for some of the products and see the product counts work.

Let's open up the development tools and have a look at the sessionStorage under the **Application** tab:

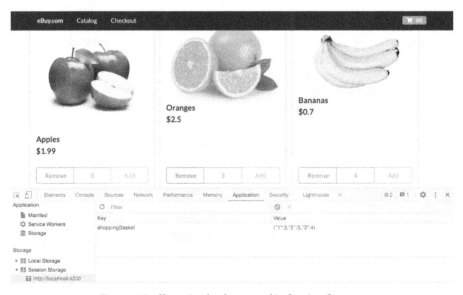

Figure 4.6 – Shopping basket stored in Session Storage

Once the state is present in **Session Storage**, we will need to read it from multiple places across different components. It is best to create it as a utility function that can be reused as needed.

We will create another library using Nx Console, but this time instead of creating a React library, we will use the `@nrwl/workspace - library` template to generate our generic `utils` library and use the import scope called `@ebuy/utils`.

The information we fill in during the `Nx Console > generate` step is as follows:

- **@nwrl/workspace**: `library`
- **Name**: `utils`
- **importScope**: `@ebuy/utils`

Running this command will generate the `utils` folder and also create the `utils.ts` file. Let us rename it to `get-session-storage.ts`.

Add the following code to read the values of a given key:

```
export function getSessionStorage(key: any) {
  const sessionStorageValue = JSON.parse(
    window.sessionStorage.getItem(key) || '{}'
  );
  return sessionStoragevalue;
}
export default getSessionStorage;
```

As you can see, this is a very simple function that accepts a key and returns the values from session storage for the given key.

Next, we will get the mini basket in the header hooked up to show the total items in the shopping basket. In the `header.tsx` file, let us add the necessary code to read and total up the items in the shopping basket.

Let us import the necessary functions:

```
import { useEffect, useState } from 'react';
import { useEventListener } from 'usehooks-ts';
import { getSessionStorage } from '@ebuy/utils';
```

We will create our function to calculate the total count like so:

```
const getTotalBasketCount = (basket: any): any => {
  return Object.values(basket).reduce((a: any, b: any) => a +
b, 0);
};
```

Next, within the `Header` component, we will use a combination of `useEffects` and `eventListeners` to ensure that the mini basket updates every time items are added to or removed from the cart:

```
const [miniBasketCount, setMiniBasketCount] = useState(null);

  useEffect(() => {
    const basket: any = getSessionStorage('shoppingBasket');
    const totalCount: any = getTotalBasketCount(basket);
    setMiniBasketCount(totalCount);
  }, []);
  useEventListener('session-storage', () => {
    const basket: any = getSessionStorage('shoppingBasket');
    const totalCount: any = getTotalBasketCount(basket);
    setMiniBasketCount(totalCount);
  });
```

Finally, we will update the shopping cart icon to display {miniBasketCount} like so:

```
<Menu.Item position="right">
        <Label>
          <Icon name="shopping cart" />
          {miniBasketCount}
        </Label>
      </Menu.Item>
```

Run the apps and try adding and removing items using the **Add** and **Remove** buttons, and see how the counts update.

The last part of this chapter is where we will complete the shopping cart component in the checkout app.

What we need to do is fetch the data for the `shoppingBasket` key in `sessionStorage` and display the products and the quantity added to the cart.

We open the `app.tsx` checkout file located in the `apps/checkout/src/app/app.tsx` file and follow these steps to get the data from `sessionStorage`:

First we import `getSessionStorage` like so:

```
import { getSessionStorage } from '@ebuy/utils';
```

Then within the App function we add the following:

```
const basketFromStorage: any =
getSessionStorage('shoppingBasket');
    console.log('Basket: ', basketFromStorage);
```

When we run the app and have a look at the console, we will be able to see the array of items from shoppingBasket.

Since shoppingBasket only stored the product IDs and their quantity we will need to map the product IDs to the product names so that we can display names in the shopping basket.

Let us create another function to do that. We will call it createCompleteBasket:

```
const createCompleteBasket = (allItems: any, quantities: any)
=> {
  return allItems
    .filter((item: any) => quantities[item.id])
    .map((item: any) => {
      return {
        ...item,
        quantity: quantities[item.id],
      };
    });
};
```

And then, finally, within our app's component's function, we create completeBasket by filtering and mapping the values from the product list to shoppingbasket like so:

```
const completeBasket = createCompleteBasket(
    PRODUCT_LIST_MOCKS,
    basketFromStorage
  );
```

Now we update the ShoppingBasket component to pass in this new prop like so:

```
<ShoppingBasket basketList={completeBasket} />
```

Test your app in the browser and give it a play. Add and remove items to and from the basket in the catalog app and then navigate to the checkout app to see the shopping basket all synced up and displaying the correct list of items.

A note on the coding samples

As you must have seen, in numerous places we have used the 'any' type definition and have skipped a few details (including unit tests). This is intentional to avoid overcomplicating the examples so that we stay focused on the core aspects of this chapter, such as routing between apps and sharing state. When building an app for production, we would encourage you to define the correct types and interfaces to take advantage of the full power of TypeScript and write relevant tests.

With this, we come to the end of this rather intense section... Take a break. Well done!

We covered a lot here. We picked up from where we had left off in the previous chapter and added a shared header component to our apps. We then set up routing via a proxy so that we could navigate between the two different apps, but as if they were part of the same domain and port. We also saw how to share state between the two mini-apps using session storage. We then created a common custom hook to store and retrieve data from session storage, and while doing so we built up the bare bones of an e-commerce app, adding items to the cart and updating the cart information on the checkout app and the mini cart on the header.

Summary

This was a long chapter, so well done for staying with us until the end. We started off by looking at what the multi-SPA pattern looks like. We saw how this pattern would be most suitable for very large applications such as a banking portal, a government portal, or an e-commerce site. We saw the architecture pattern where all these different mini-apps can take advantage of a shared common library of components and utilities to ensure the consistency of the different apps.

We then took a deep dive into code and went about setting up our two mini-apps within the Nx monorepo, after which we went about creating our shared UI header component and used Semantic UI to build out our catalog and checkout apps. This was also a good opportunity for us to see how to use scoped names, which makes our import paths look clean and simple.

Then we went about setting up the routing so that we could navigate between the two different apps, and finally, we set up a custom hook to store our app state in session storage and saw how to have it synced between the two mini-apps.

In the next chapter, we will look at the micro-apps pattern where we will have multiple micro-apps loaded within the same page.

Implementing the Micro-Apps Pattern for Microfrontends

In the previous chapter, we saw the multi-SPA pattern for building microfrontends, which is ideal for building large-scale applications where each SPA contains its own user journey.

The primary advantage of this pattern is that each app is completely independent of the others, and they are connected via a namespaced primary route that is external to the app. As a user, while navigating through the app, you will have also noticed that when you move from one SPA to another, you are redirected via the browser and the page reloads. If this is something you'd like to avoid, and if you want a consistent SPA experience, then we can explore the micro-apps-pattern that uses Module Federation.

In this chapter, we will go about building a micro-apps microfrontend, where we will learn about the following:

- What is Module Federation, and why is it a key to building microfrontends?
- Setting up a microfrontend app with host and remote apps
- Breaking down the app into smaller micro apps that are loaded via Module Federation
- Setting up routing between the different pages
- Sharing state between the different micro apps

By the end of this chapter, we will have converted our multi-SPA microfrontend into a micro-apps microfrontend using Module Federation. In doing so, we will have also learned about Zustand, an easy-to-use state management library.

Technical requirements

As we go through the code examples in this chapter, we will need the following:

- A PC, Mac, or Linux desktop or laptop with at least 8 GB of RAM (16 GB preferred)
- An Intel chipset i5+ or a Mac M1 + chipset
- At least 256 GB of free hard disk storage

You will also need the following software installed on your computer:

- Node.js version 18+ (use nvm to manage different versions of Node.js if you have to)
- Terminal: iTerm2 with OhMyZsh (you will thank me later)
- IDE: We strongly recommend VS Code as we will be making use of some of the plugins that come with it for an improved developer experience
- NPM, Yarn, or PNPM. We recommend PNPM because it's fast and storage-efficient
- Browser: Chrome, Microsoft Edge, or Firefox
- A basic understanding of Nx.dev monorepos and a basic understanding of using the NX console plugin in VS Code
- Working knowledge of React

The code files for this chapter can be found here: `https://github.com/PacktPublishing/ Building-Micro-Frontends-with-React`.

We also assume you have a basic working knowledge of Git, including branching, committing code, and raising a pull request.

Why do we need Module Federation for Microfrontends?

In the multi-SPA approach to microfrontends, you may have noticed that we end up duplicating some of the common dependencies across the different micro apps. In the grand scheme of things, when the primary goal is to keep things simple, this would be an acceptable trade-off. However, when the number of dependencies being duplicated and the number of apps being built are high, you need to optimize things and minimize duplication. Trying to achieve this before Webpack 5 would have led to having to deal with complex dependency management. It would also have made it difficult to maintain and evolve microfrontend applications. Module Federation helps us solve these challenges.

In the next sections, we will learn more about what Module Federation is and how it helps with building microfrontends.

What is Module Federation?

Module Federation is a new feature introduced in Webpack 5 that allows us to load external JS bundles in real time.

Before Module Federation, the standard way to import all the necessary modules for an application was only during build time, where it created a large JS bundle or smaller chunks that got loaded based on page routes, but it wasn't quite possible to dynamically load an app bundle in real time.

Module Federation provides us with a radically new way to architect our apps, build and deploy shared components, and update them without the need to rebuild the entire application.

Traditionally, we build most of our shared components, such as UI component libraries or npm modules, and import them into our application during build time. With Module Federation, these modules can be hosted at an independent URL and imported into the application at runtime. We take advantage of this very same feature to build our microfrontend architecture, where we have our micro apps independently hosted and loaded into the host or shell app in real time.

Before we get into how to go about doing it, let us look at some basic terminology associated with Module Federation. Module Federation revolves around a few concepts. Here are some of them.

ModuleFederationPlugin

All of Module Federation's features are made available in Webpack 5+ via the `ModuleFederationPlugin` plugin. This is where you define the settings of how Module Federation should work.

This plugin allows a build to provide or consume modules during **runtime** with other independent builds.

You can read in detail about `ModuleFederationPlugin` and its specs here: `https://webpack.js.org/plugins/module-federation-plugin/`.

In its simplest form, the code for `ModuleFederationPlugin` should look like this:

```
const { ModuleFederationPlugin } = require('webpack').container;
module.exports = {
  plugins: [
    new ModuleFederationPlugin({
      // module federation configuration options
    }),
  ],
};
```

The preceding code is the skeleton that holds all the configurations required to enable Module Federation.

Host apps

This is the root application within which *remote* or external apps are loaded. The host app's Module Federation configuration stores the list of remote apps that need to load within it. In our use cases of microfrontends, the host app also contains information about the different routes and the mapping of the routes to the respective remote apps.

Webpack's configuration for Module Federation in the host app should look like this:

```
module.exports = {
  plugins: [
    new ModuleFederationPlugin({
      name: 'hostAppName',
      remotes: {
        app1: '<app1's URL path to remoteEntry.js>',
        app2: '<app2's URL path to remoteEntry.js>',
      },
    }),
  ],
};
```

The preceding code is simple to understand. We let Module Federation know the name of the host app and provide a list of remote apps and the path to their corresponding `remoteEntry` file in the `remotes` object.

Remote Apps

Remote apps, as you would have guessed, are apps that load dynamically within the host app. These remote apps are also referred to as *containers* in Module Federation terminology. The JS bundle of these remote apps is usually exposed via a single `.js` file usually called `remoteEntry.js`, which the host app looks out for.

Every remote app is exposed in Webpack's Module Federation configuration in the following way:

```
new ModuleFederationPlugin({
    name: 'remoteAppName', // this name needs to match with the
entry name
    exposes: ['./public-path/remoteEntry.js'],
    // ...
  }),
```

Every remote app needs to have a unique name defined in its `name` property, and this name needs to match the names that are part of the `remotes` object defined in the host app's Module Federation configuration.

remoteEntry.js

The `remoteEntry.js` file is a small JS file that is created by Module Federation at runtime. It contains metadata for each of the remote apps. The host app relies on the `remoteEntry.js` file to know which modules to load into.

Use cases for Module Federation are not just limited to building microfrontends; they can also be used to dynamically load common libraries or a module such as a design system, for example, negating the need to publish these common libraries as npm packages and having to rebuild and re-deploy every time a common library has changed.

The following diagram helps to explain how Module Federation works:

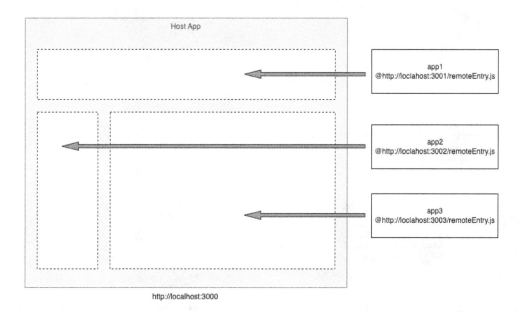

Figure 5.1 – Module Federation with three micro apps loaded in real time

From the diagram, we see that we have apps running on ports 3001, 3002, and 3003. Each of them has its metadata exposed in its respective `remoteEntry.js` files. These apps are dynamically loaded into the host app that is running on port 3000 via Module Federation.

It may be prudent to know that it is not just apps. Any kind of JS module can be dynamically imported into Module Federation.

In the next section, we will put all of this into practice.

Setting up Microfrontends with a Host and Remote app

We are going to take our multi-SPA app and convert it into a microfrontend with a Host and Remote app using Module Federation. As mentioned earlier, the main benefit of this approach is that users get a true single-page experience while still ensuring that each app is independently built and deployed.

Let us see what it takes for us to do this:

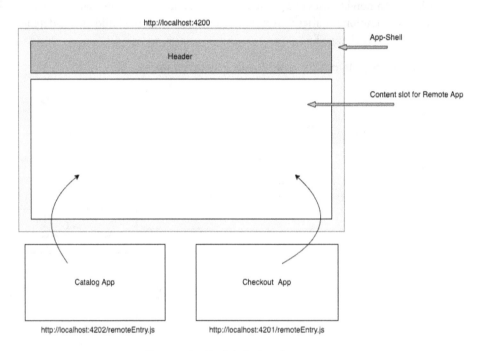

Figure 5.2 – Module Federation setup

You will notice that *Figure 5.2* is similar to *Figure 5.1* and explains the implementation details of Module Federation. We see that the Host app contains the header component and runs on port 4200. We then have our Catalog and Checkout apps running on ports 4202 and 4201. We aim to load these remote apps dynamically whenever the correct route is called.

To convert our multi-SPA into a module-federated microfrontend, we will need to make the following changes:

1. Create a new Host app that we'll call App-shell.
2. Remove the header component from each SPA and move it into App-shell.
3. Define the remote apps, namely Catalog and Checkout, that need to be loaded into the host app.
4. Define the remote entry for the Catalog and Checkout micro apps.

Let us get started. Open up the e-buy app that you built in the previous chapter.

You may also download it from the Git repo:

`https://github.com/PacktPublishing/Building-Micro-Frontends-with-React/tree/main/ch4/ebuy`

In the coming subsections, we will see how to create our host and remote apps, but first, we will clean up our existing apps and prep them to use Module Federation.

Clean up

With Module Federation, the host app takes care of routing, and there is no need for us to use the proxy configurations we defined in the `proxy.conf.json` file. So, we will delete this file and remove the unnecessary configuration from the `project.json` file.

Go ahead and delete `/apps/catalog/proxy.conf.json` and, in the `catalog/project.json` file, delete the following line:

```
"proxyConfig": "apps/catalog/proxy.conf.json"
```

While we're at it, we can also get rid of `baseRef`, which we defined in our `checkout/project.json` file. Locate this line and delete it:

```
"baseHref": "/checkout/"
```

Setting up the App-shell host app

With this, we are now set to start migrating our multi-SPA apps to Module Federation.

Nx Console has a nifty generator for creating a host and remote apps for Module Federation. Follow these steps:

1. Create a React host app:

    Nx Console | Generate | @nrwl/react – host Generate a host react application

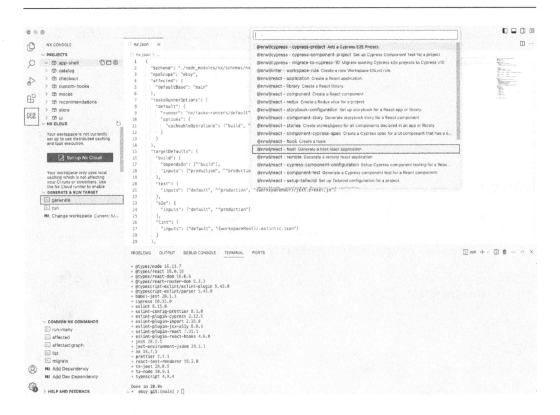

Figure 5.3 – Selecting the host app generator from Nx Dev Console

2. Enter the following information in this form:

    • **Name**: app-shell

    • **devServerPort**: 4200

    • **e2eTestrunner**: none

    • **remotes**: We will leave this blank and add it manually due to a bug that doesn't allow multiple app names

3. After you hit the **Generate** button, you will see the app-shell folder within apps.

    When you navigate into the folder and have a peek at the files in it, you will notice that it has a new file on the root of the app called apps/app-shell/module-federation. config.js.

4. Open up the file and, in the remotes array, add Catalog and Checkout as the remote apps:

    ```
    remotes: ['catalog', 'checkout']
    ```

5. Now let us open up the `apps/app-shell/src/app/app.tsx` file and have a look.

    You will notice that this uses familiar React concepts such as `React.Suspence` and React Router's `Route`.

6. We will tweak this boilerplate file:

```
import React from 'react';
import { Container } from 'semantic-ui-react';
import { Route, Routes } from 'react-router-dom';
import 'semantic-ui-css/semantic.min.css';
import { Header } from '@ebuy/ui';

const Catalog = React.lazy(() => import('catalog/Module'));
const Checkout = React.lazy(() => import('checkout/Module'));

export function App() {
  return (
    <React.Suspense fallback={null}>
      <Container style={{ marginTop: '5rem' }}>
        <Header />
        <Routes>
          <Route path="/" element={<Catalog />} />
          <Route path="/catalog" element={<Catalog />} />
          <Route path="/checkout" element={<Checkout />} />
        </Routes>
      </Container>
    </React.Suspense>
  );
}

export default App;
```

    As you can see from the preceding code, we first import the `Header` component into App-shell.

    You will also notice that we are using dynamic imports to import our Catalog and Checkout apps using `React.lazy`. These lines will currently be throwing errors as it is unable to find the module.

7. To solve this, create a file called `/apps/app-shell/src/remotes.d.ts` with the following code:

```
declare module 'catalog/Module';
declare module 'checkout/Module';
```

    The `remotes.d.ts` file is used to provide TypeScript type declarations for remotes in a Module Federation setup.

Further down in the JSX, you will notice that we import the Catalog app on the / and / catalog routes while we import the Checkout app on the /checkout route.

This more or less completes the setup for the host app.

Setting up our Remote apps

Setting up our remote apps will take a bit of work. Let us crack on and work on them one at a time.

Here's what we need to do in order to convert an existing React app within Nx to a remote app:

- Create remote entries in the module-federation.config.js file.

- Swap the app builder in project.json to use the module-federation plugin.

- Add a serve-static executor.

- Use a custom Webpack configuration that defines the remote entry modules.

Let us carry out the preceding changes in the Catalog app to start with. Follow these steps:

1. In the apps/catalog folder, create a new file called module-federation.config. js and add the following code:

```
const moduleFederationConfig = {
  name: 'catalog',
  exposes: {
    './Module': './src/app/app.tsx',
  },
};

module.exports = moduleFederationConfig;
```

This is where we define the Catalog remote app and the module path that it exposes.

2. Next, we need to make a couple of changes to the apps/catalog/project.json file.

3. First, we add a new command under the targets and call it serve-static:

```
"serve-static": {
  "executor": "@nrwl/web:file-server",
  "defaultConfiguration": "development",
  "options": {
    "buildTarget": "catalog:build",
    "port": 4201
  }
}
```

Notice that we intend to run our app on 4201, so let's also make sure the serve command also uses port 4201.

4. Make sure the port under the regular `serve` command is defined within the `options` object:

```
"serve": {
      "executor": "@nrwl/web:dev-server",
      "defaultConfiguration": "development",
      "options": {
        "buildTarget": "catalog:build",
        "hmr": true,
        "port": 4201
      },
```

This is because the Module Federation plugin expects the port to be defined within the `options` object. If not, it will use a default port, which can lead to very interesting bugs.

Refer to this line in the source code: `https://github.com/nrwl/nx/blob/master/packages/react/src/module-federation/with-module-federation.ts#L29`.

5. Next, under the `serve` object, we update the executor to use `module-federation dev-server`:

```
"serve": {
      "executor": "@nrwl/react:module-federation-dev-server",
```

6. Next, ensure we have `WebpackConfig` with a custom Webpack configuration:

```
"webpackConfig": "apps/catalog/webpack.config.js"
```

7. Now let us update `webpack.config.js` with the following code:

```
const { withModuleFederation } = require('@nrwl/react/module-
federation');
const baseConfig = require('./module-federation.config');

const defaultConfig = {
  ...baseConfig,
};

module.exports = withModuleFederation(defaultConfig);
```

Now let us repeat the same steps for the Checkout app:

1. In the `apps/checkout/` folder, create a new file called `module-federation.config.js` with the following code:

```
const moduleFederationConfig = {
  name: 'checkout',
  exposes: {
    './Module': './src/app/app.tsx',
```

```
    },
  };

  module.exports = moduleFederationConfig;
```

As you can see, it is identical to what we had on the Catalog app. The only difference is that we changed the name value to checkout.

2. Next, let us add the serve-static command to the targets object in the apps/checkout/project.json file:

```
"serve-static": {
"executor": "@nrwl/web:file-server",
"defaultConfiguration": "development",
"options": {
  "buildTarget": "checkout:build",
  "port": 4202
}
}
```

3. In the same file, we continue to update the executor:

```
"serve": {
    "executor": "@nrwl/react:module-federation-dev-server",
```

4. Then under the serve.options update the port number to 4202.

5. We also update webpackConfig:

```
"webpackConfig": "apps/checkout/webpack.config.js"
```

Since there are no changes to the webpack.config.js file, we can simply copy and paste this file from the Catalog app.

6. Finally, we will update the Header component to use the Link component from ReactRouter so that we get that single-page experience.

7. Open up the /libs/ui/src/lib/header.tsx file and update the following to use <Link> instead of <a>:

```
<Link to={navItem.href ?? '#'}>{navItem.label}</Link>
```

8. Don't forget to import the <Link> command:

```
import { Link } from 'react-router-dom';
```

9. Before we try testing, let us not forget to remove the header component from the respective Catalog app located at /apps/catalog/src/app/app.tsx and the Checkout app at /apps/checkout/src/app/app.tsx.

10. Let's do a quick test on the terminal. Run the following command:

```
pnpm nx serve app-shell
```

As you see, the commands being executed in the terminal notice that the Catalog app is being built along with the `app-shell serve` command.

11. Once everything is running without any errors, open up `http://localhost:4200` and verify that the Catalog and Checkout apps load up on the correct routes.

    You will also notice that the product images don't show up any longer. This is because the app is looking for images in the `/assets` folder of the App-shell app.

12. In the multi-SPA approach, the Catalog app was the default route and was sort of acting like the host app. Since App-shell is now our host, we will need to copy the images from the `/catalog/src/assets` folder into the `app-shell/src/assets` folder. Once you have done this, the images should load up into the app.

13. Navigate between the Catalog and Checkout apps. Add items to your cart and enjoy seeing the apps work nicely.

    Since everything is going well, and since each micro app team should be able to work on their individual apps, let us also make sure that we can run each app individually.

14. Run `pnpm nx serve catalog` and you'll notice you get an error:

```
Error
Shared module is not available for eager consumption: webpack/
sharing/consume/default/react/react
```

    This is due to Module Federation treating the Catalog app as a bidirectional host and not being able to eagerly load the shared modules.

    You can read more about it here:

    `https://webpack.js.org/concepts/module-federation/#uncaught-error-shared-module-is-not-available-for-eager-consumption`

    To overcome this issue, we need to define an asynchronous boundary to split out the initialization code of a larger chunk and avoid any additional roundtrips to the server.

15. To solve it, we need to make a couple of tweaks. In the Catalog app, let us first rename `/apps/catalog/src/main.tsx` to `bootstrap.tsx`.

16. Next, we create a new file called `main.ts` within the same `src` folder and have a single line importing bootstrap:

```
import('./bootstrap');
```

17. Next, we need to ensure that this newly created `main.ts` file is what is being used as the entry point, so now, in our `project.json` file for the Catalog app, we update the `main` property within the `build > options` object:

```
"main": "apps/catalog/src/main.ts",
```

18. Repeat the same steps for the Checkout app. Now, you should be able to run the apps as a module-federated microfrontend or each app individually.

You may have also noticed that at the start of this chapter, we referred to the `remoteEntry.js` file as the entry file for the remote apps and that we didn't really define one.

However, if you look at your dev tools' network tab, you will notice there are two `remoteEntry.js` files being called from ports `4201` and `4202` respectively. This is Nx and Module Federation doing a bit of magic here.

> **Important note**
>
> If you dig into the source code in this file, you will notice the filename being defined as part of the `ModuleFederationPlugin` configuration (`https://github.com/nrwl/nx/blob/master/packages/react/src/module-federation/with-module-federation.ts`).

The screenshot in *Figure 5.4* shows the `remoteEntry.js` file being called from the respective apps:

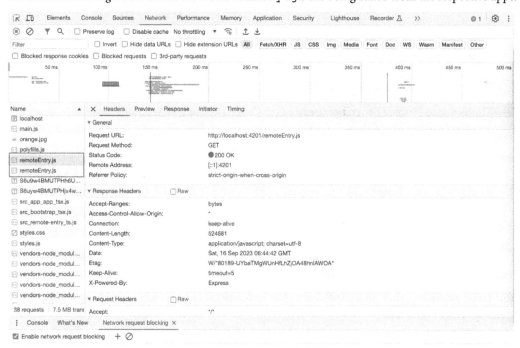

Figure 5.4 – RemoteEntry.js

If you are keen to explicitly define the file and stay as close as possible to the native workings of Module Federation, then go ahead and create a file in `apps/catalog/src/remote-entry.js` with the following line:

```
export { default } from './app/app';
```

Update the `exposes` value in the `apps/catalog/module-federation.js` file to read as follows:

```
const moduleFederationConfig = {
  name: 'catalog',
  exposes: {
    './Module': './src/remote-entry.ts',
  },
};
```

With that, we have completed the section on using Module Federation and successfully converted our multi-SPA app into a module-federated microfrontend.

In this section, we saw what minimal steps are required to get Module Federation working and what extra steps, such as defining the remotes and exposing the module names, are needed to allow each app to work independently.

In the next section, we will see how to further break down a remote app into a true micro-app microfrontend.

Extending Module Federation to a true Micro-apps Pattern

Imagine you are part of a team that manages a very large e-Commerce app (think of Amazon.com). For such large sites, it is a common practice to have teams that own a single organism-level component (`https://atomicdesign.bradfrost.com/chapter-2/#organisms`) instead of the entire mini app.

For example, we have a dedicated team that works exclusively on the Product Recommendations component. This component is injected into, say, the Catalog app.

In such a case, it would be prudent to create another micro app called Recommendations and dynamically import it into the Catalog app. This would allow for true, federated, micro-app pattern architecture.

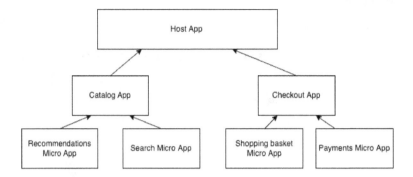

Figure 5.5 – Tree of remote apps with Module Federation

As you can see from the preceding diagram, we can further break down our Catalog and Checkout apps into smaller organism-level components and have each of them load into the Catalog app remotely via Module Federation.

> **Important note**
>
> Remember, while this may seem very cool, it is important that we don't overdo it by converting every single organism into a module-federated micro app. It is important to follow the principles of microfrontends mentioned in *Chapter 2*, namely, *breaking down the app into the largest independently deployable app owned by a single team* and not necessarily the smallest.

Having said that, and assuming you do have an independent team owning the Recommendations micro app, let us go about creating the micro app.

Creating the Recommendations Remote Micro app

Let us use our trusted Nx dev console and the GENERATE command and follow these steps:

1. Select `@nrwl/react - remote Generate a remote application` and use the following information while leaving the rest as their defaults:

    -   **Name**: `recommendations`

    -   **e2eTestRunner**: `none`

    -   **host**: `catalog`

    -   **devServerPort**: `4203`

2. Use the **Generate** command and verify that the `recommendations` app has been successfully created.

3. Let us quickly edit `apps/recommendations/src/app.tsx` to remove the boilerplate code and leave it with a simple message:

```
import 'semantic-ui-css/semantic.min.css';
export function App() {
  return (
    <div className="ui raised segment">
      <h1>Recommendations</h1>
      <p>Recommendations goes here</p>
    </div>
  );
}

export default App;
```

> **Note**
> Building a full-fledged Recommendations micro app is beyond the scope of this book.

4. Run `npx nx serve recommendations` and verify that the app loads up properly on port `4203`.

Keep it running while we work on adding it as a remote to our Catalog app.

Adding Recommendations as a Remote app to Catalog

Since we want the Recommendations micro app to load within Catalog, as a remote app, we need to convert Catalog to behave like a host. We do this using the following steps:

1. Open up the `apps/catalog/module-federation.config.js` file and add the `remotes` entry to it:

```
const moduleFederationConfig = {
  name: 'catalog',
  remotes: ['recommendations'],
  exposes: {
    './Module': './src/app/app.tsx',
  },
};

module.exports = moduleFederationConfig;
```

2.   Next, let us create a `remotes.d.ts` file within the `apps/catalog/src` folder using the following line:

```
declare module 'recommendations/Module';
```

3.   Finally, let us import and call the Recommendations app into our `apps/catalog/src/app/app.tsx` file:

```
. . .
import React from 'react';
const Recommendations = React.lazy(() =>
import('recommendations/Module'));
```

4.   In the `jsx` part of the component, here is how we call our Recommendations component:

```
<Recommendations />
```

5.   Open up a new terminal window and run the following command:

```
pnpm nx serve app-shell
```

If everything goes as planned, you will see the Catalog app with the Recommendations component loaded within it. With this, we come to the end of this section.

In this section, we saw how we can use Module Federation to further break down a host app into smaller micro apps and have them all working together as a tree of remote apps.

In the next section, we will see how to set up state management within our micro apps microfrontend.

State management with Module Federation

As you may have noticed by now, our custom state management system, which uses `sessionStorage`, continues to work seamlessly with Module Federation. This is because, from a React perspective, it all looks like a regular React application, with modules being lazy-loaded. So, one of the benefits of Module Federation is that we can use any of the regular state management concepts, such as prop drilling, context API, or libraries such as Redux or Zustand, to manage the state.

In this section, we will make use of the Zustand state management library as it is extremely user-friendly and has zero boilerplate code.

Now, logically, especially for those who use a lot of context API, we would be inclined to have the store within App-shell and have the other micro apps consume it. However, with Module Federation, this would not be ideal, because the store would need to be exposed as a remote app and imported into other micro apps that act as hosts. If you try plotting this, it kind of feels like a cyclic dependency of sorts, where App-shell acts as a host for all other components but the store located in it is a remote for other components.

The following diagram better illustrates the problem of this cyclic flow:

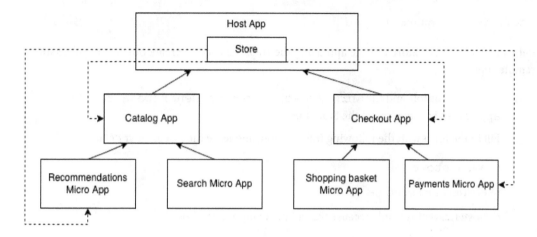

Figure 5.6 – Cyclic flow between App-shell and the store

When working with Module Federation, it is preferable to have a unidirectional flow of how the remote and host apps are loaded in. With that in mind, it would be more prudent to have our store as its own independent micro app and have it defined as a remote app to all the other apps that consume it. With this new structure, the diagram in *Figure 5.6* can be redrawn:

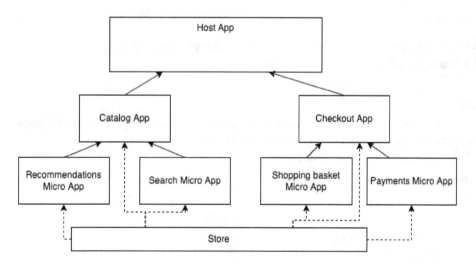

Figure 5.7 – Unidirectional remotes for the Store app

As evident from *Figure 5.7*, the unidirectional flow of remotes for the Store app looks a lot cleaner, and it ensures that App-shell isn't getting unnecessarily bloated with business logic and state.

Since we are going to use Zustand for state management, this would be a good time for us to install it. Run the following command:

```
pnpm install zustand
```

Let us now create our Store remote app using the steps we used to create our Recommendations remote app:

1. Using Nx Console and the @nrwl/react - remote Generate a remote application file, create the Store micro app.

2. Fill in the form with the following information, and leave the rest as their defaults:

    - **Name**: store

    - **e2eTestRunner**: none

    - **host**: (leave this blank because we will manually add the hosts)

    - **devServerPort**: 4204

Once the app has been created, let us go about setting up our store. To demonstrate the working of the state and store across the different micro apps, we will have a **Like** button in the host app. Clicking it will increment the like count. We will also display the count within the Recommendations app. Then, we will have a **Reset** button in the Recommendations micro app that will reset the store and verify that the like count has reset in all places.

Let us get started:

1. Navigate to the /apps/store/src folder and create a new file called store.tsx. This is where we will define our store and hooks.

> **Note**
>
> Zustand is super easy to work with. Have a look at the documentation at https://docs.pmnd.rs/zustand/getting-started/introduction.

2. Begin in the store.tsx file by importing Zustand and defining the LikeCount interface:

    ```
    import {create} from 'zustand';
    interface LikeCount {
      count: number;
      increment: () => void;
      reset: () => void;
    }
    ```

3. Next, we create our `useStore` hook and define the `initial state`, `increment`, and `reset` functions. This is the standard way to do so:

```
const useStore = create<LikeCount>((set) => ({
  count: 0,
  increment: () => set((state) => ({ count: state.count + 1 })),
  reset: () => set(() => ({ count: 0 })),
}));

export default useStore;
```

And that is it! Our store with the `useStore` hook is ready to be consumed.

4. Next, we need to expose this as a remote app. We will do this by making two additional changes. In the `apps/store/src/remote-entry.ts` file, modify the following line to the following text:

```
export { default } from './store';
```

5. Next, we let App-shell and the Recommendations app know that they need to use Store as a remote app. We do this by adding Store to the remote array in the respective `module-federation.config.js` files in App-shell and the Recommendations app:

```
//apps/app-shell/module-federation.config.js
const moduleFederationConfig = {
  name: 'app-shell',
  remotes: ['catalog', 'checkout', 'store'],
};

module.exports = moduleFederationConfig;
```

Here is what we have in the `apps/recommendations/module-federation.config.js` file:

```
const moduleFederationConfig = {
  name: 'recommendations',
  remotes: ['store'],
  exposes: {
    './Module': './src/remote-entry.ts',
  },
};
module.exports = moduleFederationConfig;
```

6. The next thing we need to do is declare the `store` module in the `remotes.d.ts` file app-shell file:

```
declare module 'store/Module';
```

7. We will need to do the same in the Recommendations app. Since the `remotes.d.ts` file doesn't exist, we can create a new file with the following line in `/apps/recommendations/src/remotes.d.ts`:

    ```
    declare module 'store/Module';
    ```

We now have the store hooked up so that Recommendations and App-shell can read and write to our Store micro app.

Adding the Like button to the host app

In this section, we will create a **Like** button that increments the like count and gets stored in the store.

Now that we have set up the remotes, let us import the store into our app shell and create the **Like** button in the `/apps/app-shell/src/app/app.tsx` file. Follow these steps:

1. Import `useStore`:

    ```
    import { Button } from 'semantic-ui-react'
    import useStore from 'store/Module';
    ```

2. Then, de-structure the count and increment within the App function:

    ```
    const { count, increment } = useStore();
    ```

3. Finally, in our JSX, we add our button after the `<Header/>` component:

    ```
    <Button onClick={increment}>{count} Likes </Button>
    ```

4. Restart all the apps. You can also make use of the following custom command that we created to serve all apps:

    pnpm serve:all

We've got the state working with Zustand within the host app, and as you can tell, it is refreshingly simple and devoid of any boilerplate code. But the true purpose of us setting up the state and store is to ensure that this state can be shared with other micro apps. In our case, it will be the Recommendations app, right at the bottom of the federation hierarchy.

5. In the `apps/recommendations/src/app/app.tsx` file, our code should look very similar to the following:

    ```
    import 'semantic-ui-css/semantic.min.css';
    import { Button } from 'semantic-ui-react'
    import useStore from 'store/Module';
    export function App() {
      const { count, reset } = useStore();
      return (
        <div className="ui raised segment">
    ```

```
          <h1>Recommendations</h1>
          <p>Recommendations goes here</p>
          <p> {count} people liked the recommendations</p>
          <Button onClick={reset}>reset</Button>
        </div>
    );
}


export default App;
```

That's it!

6. Run your apps and play around with the **Like** button. Reset it and verify that the count stays
    in sync between the host and the Recommendations app:

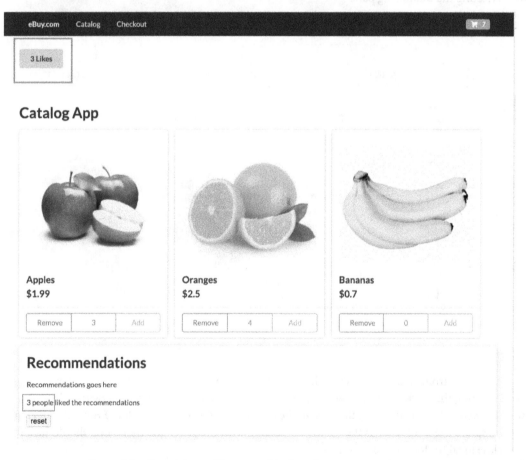

Figure 5.8 – Complete working app with shared state across micro apps

Our app works great, but let us make sure it's performant as well!

Avoiding Unnecessary Re-rendering

When working with the state, a very important performance-related point to check is avoiding unnecessary re-rendering. This is especially true when the state is being shared between different components or when it is being prop drilled.

One way to verify that is to go into **Developer Tools**, open up the **Rendering** pane, select **Paint flashing** and **Frame Rendering Stats**, and verify that when you click on the buttons, only the necessary items within the components are updating.

In Chrome, you can access this panel by opening up **Developer Tools**, going into **More Tools**, and then selecting the **Rendering** pane:

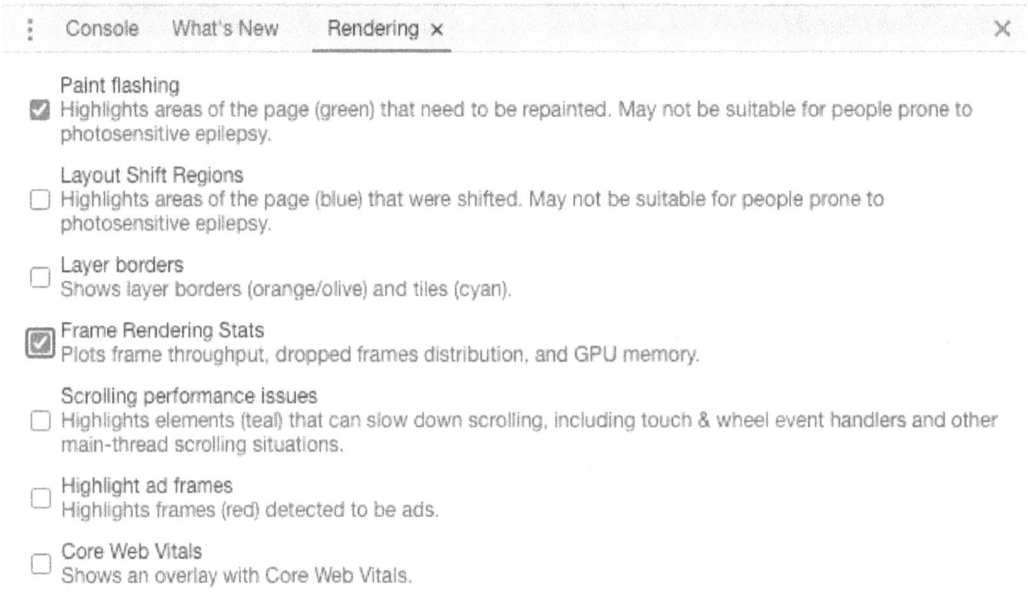

Console   What's New   Rendering ×

☑ Paint flashing
Highlights areas of the page (green) that need to be repainted. May not be suitable for people prone to photosensitive epilepsy.

☐ Layout Shift Regions
Highlights areas of the page (blue) that were shifted. May not be suitable for people prone to photosensitive epilepsy.

☐ Layer borders
Shows layer borders (orange/olive) and tiles (cyan).

☑ Frame Rendering Stats
Plots frame throughput, dropped frames distribution, and GPU memory.

☐ Scrolling performance issues
Highlights elements (teal) that can slow down scrolling, including touch & wheel event handlers and other main-thread scrolling situations.

☐ Highlight ad frames
Highlights frames (red) detected to be ads.

☐ Core Web Vitals
Shows an overlay with Core Web Vitals.

Figure 5.9 – Rendering pane in the developer consoles

As you can see from *Figure 5.9*, once we have **Paint flashing** enabled, you will see a green rectangle encapsulating the sections of the page that re-render due to a change or user interaction. The ideal state is when only a small part of the page flashes when the user interacts with it. **Frame Rendering Stats** displays the frame rate as the user is interacting with the page. The frame rate should ideally stay close to 60 fps for a smooth user experience.

And with that, we come to the end of this section regarding state management with Zustand in a module-federated application. In this section, we learned the benefits of defining the Store app as a separate micro app. It is then dynamically imported into the other micro apps. We learned how to go about setting up the store as a module-federated module. We then saw how the host app and the Recommendations app can share state via the shared store. Finally, we were also able to turn on paint flashing and frame rendering to verify that as the state changes, only the necessary elements within the apps update, and it doesn't cause components that haven't changed to be re-rendered.

Summary

We've finally come to the end of another interesting chapter. We started by learning about Module Federation and how it is a game-changer in the way we build and maintain apps. We learned some of the basic concepts of Module Federation, such as host apps, remote apps, `remoteEntry`, and more.

We then saw how to convert our multi-SPA app into a module-federated app with an app shell and how to load the Catalog and Checkout apps as remote apps. Then, we further broke things down to include smaller micro apps within these apps to create a tree of module-federated micro apps. Finally, we saw some of the best practices of managing the state and saw how we can look at tools such as Zustand to manage the state between these different micro apps.

In the next chapters, we will see how to build these apps for production and how to deploy them to static storage on the cloud.

6

Server-Rendered Microfrontends

Most JavaScript frameworks, including React, are primarily used to build **client-side-rendered** (**CSR**) applications. Client-rendered apps are great for certain use cases, such as admin dashboards or banking apps where users interact with the app in a logged-in area. CSR apps are not ideal for use cases where users access a site via a search engine or for anonymous short user journeys, such as news sites, blogs, or guest checkouts on e-commerce sites. This is because many search engine bots are not capable of indexing CSR-based web apps. CSR apps also have a poor **Largest Contentful Paint** (**LCP**) score – that is, their first-time page load performance scores are bad, leading to higher bounce rates.

To overcome these drawbacks, it is now an accepted practice to have a web app's pages rendered on a Node.js server and serve the rendered HTML pages to the browser. This is commonly known as **Server-Side Rendering** (**SSR**), or a Server-Side-Rendered (SSR) app.

In this chapter, we will look at how to build a module-federated microfrontend for a server-side-rendered app. While the process for implementing module federation is very similar to what we saw in the previous chapter, the fact that the pages are server-side-rendered brings a bit of complexity, and we will look at some of the nuances that we need to deal with when it comes to implementing a microfrontend with SSR.

In this chapter, we will cover the following topics:

- A quick look at how CSR and SSR apps differ
- Learning about Next.js and Turbo repo
- Learning how to set up hosts and remote apps with Next.js and module federation
- See how to expose multiple components as remotes that can be consumed into different apps
- Looking into issues relating to hydration of state in SSRs and also how to go about reflecting the changes made in one micro app in the main app

By the end of this chapter, we will have a server-side-rendered microfrontend built using Next.js.

Technical requirements

As we go through the code examples in this chapter, we will need the following:

- A PC, Mac, or Linux desktop/laptop with at least 8 GB of RAM (16 GB is preferred)
- An Intel chipset i5+ or a Mac M1 + chipset
- At least 256 GB of free hard disk storage
- A basic understanding of Next.js and Turborepo would be ideal
- A basic understanding of Node.js would be helpful

You will also need the following software installed on your computer.

- Node.js version 18+ (use nvm to manage different versions of Node.js if you have to).
- **Terminal**: iTerm2 with OhMyZsh (you will thank me later).
- **IDE**: We strongly recommend VS Code, as we will make use of some of the plugins that come with it for an improved developer experience.
- NPM, Yarn, or PNPM; we recommend PNPM because it's fast and storage-efficient.
- **Browser**: Chrome, Microsoft Edge, or Firefox.

The code files for this chapter can be found here: `https://github.com/PacktPublishing/Building-Micro-Frontends-with-React`.

We also assume you have a basic working knowledge of Git, such as branching and committing code and raising a pull request.

How do Client Rendered and Server Rendered Apps differ?

When it comes to building web apps with JavaScript, there are two primary methods in terms of how a user interface gets built and served to the user. They are referred to as **Client-Side-Rendered (CSR)** and **Server-Side-Rendered (SSR)**.

From a development standpoint, coding a CSR or an SSR app predominantly remains the same, except for some additional steps for SSR. However, there are differences in the internal working of these apps in terms of how they are rendered, and also in how they can be deployed on the cloud.

In this section, we will look a bit deeper into these differences.

Client Side Rendered Apps (CSR)

Let us have a look at how a Client Side app works. As its full name suggests, the CSR app is "rendered" on the client. In short, the app runs within the user's browser, makes a call to fetch data, and the page is generated on the browser. The following diagram illustrates this better:

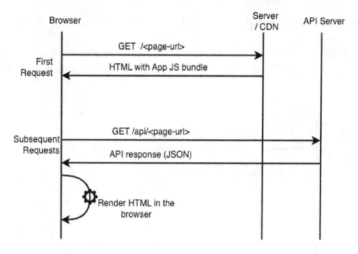

Figure 6.1 – The request and response flow for a CSR app

The preceding *Figure 6.1* illustrates the request flow in a CSR application. Here, the browser makes a first call to the server for a given URL, and the server (or sometimes the CDN itself) will respond back with a nearly empty HTML shell, containing the link to the app's JavaScript bundle. The browser parses the bundle and then makes a second AJAX call to the server API, receiving the JSON response for the given URL. The browser then parses the response and, based on the views in the client-side app, renders the HTML page in the browser before serving it to the user. For every other call, the browser continues to make AJAX calls to the API endpoint and parses the page on the browser.

With this flow, note that for the very first request from the user, there are two round trips to the server – first, to fetch the JavaScript bundle, and second, to get the page data and render the page.

Due to the nature of how CSR apps work, they are ideally suited for user experiences where users generally stay logged into an app and navigate through multiple pages per session.

Some of the drawbacks of Client side Rendered apps are as follows:

- For the very first request, users have to wait a bit longer due to the additional round trip to the server

- Since the server response doesn't contain any actual HTML data, search engine bots that are not optimized to parse JavaScript will have difficulty in indexing content from a client-rendered app

CSR apps are not suited for scenarios where the user journey is short, such as e-commerce websites where a user arrives via a search result link, buys a product or two, and leaves, or a blog site where users generally read only one to two articles at a time.

Now, let us see how an Server Side Rendered app works.

Server Side Rendered Apps (SSR)

In a Server Side Rendered app, as the full name suggests, for the very first request the page is generated on the server, and the rendered HTML page is sent to the browser. Let us look at it in a bit more in detail:

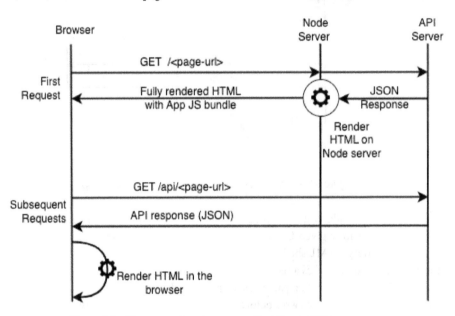

Figure 6.2 – The request and response flow for an SSR app

The working of an SSR app is illustrated in the preceding *Figure 6.2*. What we see here is when the first request for a page is made from the browser to the Node.js server, it in turn makes a call to the API server to fetch the data. Then, the HTML page is generated on the server itself and sent back to the browser, along with the initial state and the JavaScript bundles. The state hydrates on the browser, then all subsequent calls are made from the browser to the API server, and the pages are rendered on the browser itself.

Since the browser receives a fully rendered HTML page on the first request itself, the perceived performance for end users is good. It also helps with **Search Engine Optimization** (**SEO**), especially where search engine bots are not very good at parsing CSR pages.

Server-rendered apps are preferred for web apps where user journeys are short, such as B2C e-commerce apps, or content-heavy apps such as news sites or blogs.

We now have a good understanding of how SSR and CSR apps work, what their pros and cons are, and what use cases are most suited for each of them. With this information, let us start building our SSR microfrontend in the next section.

Building out our Server Rendered Microfrontend

In this section, we will look at how to build SSR apps using a meta framework such as Next.js, and then we will take it further to build a module-federated microfrontend using webpack's module federation plugin. While doing so, we will explore another monorepo tool called Turborepo.

> **Important Note**
>
> At the time of writing this book the Module Federation Plugin doesn't support Next.js 13 and the App Router and hence for this chapter we will use Next.js version 12

When it comes to building an SSR app in React, there are two common approaches:

- **A custom build using Node.js**: Here, we set up a Node.js server, render the React app on Node.js, stringify the response using the `renderToString` or `renderToPipeableStream` methods, and then use the `hydrateRoot` method, which are all part of the `react-dom/server` module to attach React to the rendered HTML

- **Use an SSR meta-framework**: Meta-frameworks such as Next.js, Remix, or Shopify's Hydrogen can abstract away all the complexities of setting up an SSR app and provide a simple interface to build performant SSR React apps

For this chapter, we will use Next.js to build our SSR app. Next.js is one of the oldest and most popular frameworks to build SSR React apps.

For the mono repo, we will use another tool called Turborepo. While we can build Next.js apps with Nx monorepos as well, we will choose Turborepo so that we can also learn about the nuances of the different monorepo tools and how they operate.

Getting started with Turborepo and Next.js

Next.js is the most popular meta-framework that allows you to build SSR apps with React. Turborepo is another new monorepo framework that is gaining popularity, and it was recently acquired by Vercel, the company that builds and maintains Next.js.

While we will cover the essentials of Turborepo and Next.js in this chapter, I strongly encourage you to spend time going through their docs to get a deeper understanding of how these frameworks work.

We will start from a clean slate here; let us begin by creating our monorepo with Turborepo:

1. Run the following command in the terminal:

    ```
    pnpx create-turbo@1.6
    ```

    Alternatively, you can run the following:

    ```
    npx create-turbo@1.6
    ```

2. This will download a bunch of libraries and then prompt you to decide where you'd like your monorepo to be created. Let's call it ebuy-ssr.

3. On the next prompt to assign a package manager, you can choose the one you prefer. For the purpose of this chapter, we will choose pnpm.

4. Let Turborepo go and do its stuff, and after the process is complete, you can cd into the ebuy-ssr folder and run the following command:

    ```
    pnpm dev
    ```

5. Note that it launches two apps, web and docs, on ports 3000 and 3001, respectively. In the browser, open up http://localhost:3000 and http://localhost:3001 and have a look at the really minimalistic default pages.

6. Open up the ebuy-ssr folder within your IDE and take a look at the folder structure.

    It will look something like this:

    ```
    .
    └── ebuy-ssr/
    ├── apps/
    │   ├── docs
    │   └── web
    ├── packages/
    │   ├── eslint-config-custom
    │   ├── tsconfig
    │   └── ui
    ├── package.json
    └── turbo.json
    ```

    The key files and folders that we need to consider are as follows:

    -   apps: This is the folder that will hold all our micro apps.
    -   packages: This is the folder where we keep all our utilities, shared components, libraries, and so on. It is the equivalent of the libs folder in Nx.
    -   package.json: The package.json files play a crucial role in how the turbo monorepo functions.
    -   turbo.json: This is the file where we define the configurations for Turborepo.

The differences between Turborepo and Nx

While both Turborepo and Nx do the same job of managing a monorepo for us, there are differences in their approach. Nx feels like a thin layer of abstraction that allows us to manage our monorepos, mainly via configurations. We tend to heavily rely on NX and its commands to build and manage our mono repos; Nx really doing all the heavy lifting for us. Turborepo, on the other hand, is quite lightweight and relies more on the npm package manager's standards to manage the monorepo. Turborepo's approach is to stay invisible in the background and let the developers have full control over how they manage their monorepos. This also means you need to do a bit more work when managing your monorepo with Turborepo.

Setting up our Micro Apps

As we can see, we have two apps created by default within our apps folder, web and docs. We will start by renaming the web folder to home let us delete the docs folder for now:

1. Rename the web folder home, delete the docs folder. Make sure that you update the name property in apps/home/package.json to "name": "home", as this is what Turborepo uses to recognize the app.

2. While we have the file open, let us define the port in which it will run in dev mode. Update the dev script in apps/home/package.json to "dev": "next dev --port 3000".

Note that with Turborepo, we have multiple package.json files. The package.json file in the root folder is used to manage the dev dependencies that are needed to manage the monorepo, and also the common dev dependencies needed for all the apps in the monorepo. We can also define our common script commands there.

The package.json file in each of the apps' folders is used to manage the workspace and the dependencies for each of the apps. The primary advantage here is that each of your micro apps has its own npm_modules folder, thereby ensuring that each team is fully independent in managing their packages and dependencies.

Creating pages and components in Next.js

Let us get started with creating a few components in our respective micro apps.

1. Creating components with Next.js is very similar to how you'd do it with other React apps; we generally create a components folder and keep our components in it.

2. When it comes to routing, Next.js 12 uses a filesystem-based router; what this means is to create a new route. We need to create a file with the route name in the /pages folder.

    For example, for a route such as http://localhost:3000/about-us, we would create a file like so – /pages/about-us.tsx.

3. Let us create our components. Since we will use `semantic-ui` to build out our components, let us go ahead and add them as dependencies in our micro apps' package managers.

4. Run `pnpm add semantic-ui-react semantic-ui-css` in `apps/home` of the micro apps folder.

5. Then, create a folder called `/components` within the home folder, and then create the `Header` component in there.

6. In the `/apps/components/Header.tsx` file, add the following code:

```tsx
import { Menu, Container, Icon, Label } from "semantic-ui-
react";
import Link from "next/link";
export function Header() {
  return (
    <Menu fixed="top" inverted>
      <Container>
        <Menu.Item as="a" header>
          eBuy.com
        </Menu.Item>
        <MenuItems />
        <Menu.Item position="right">
          <Label>
            <Icon name="shopping cart" />0
          </Label>
        </Menu.Item>
      </Container>
    </Menu>
  );
}
const MenuItems = () => {
  return (
    <>
      {NAV_ITEMS.map((navItem, index) => (
        <Menu.Item key={index}>
          <Link href={navItem.href ?? "#"}>{navItem.label}</
Link>
        </Menu.Item>
      ))}
    </>
  );
};

interface NavItem {
  label: string;
  href?: string;
}
```

```
const NAV_ITEMS: Array<NavItem> = [
  {
    label: "Catalog",
    href: "/catalog",
  },
  {
    label: "Checkout",
    href: "/checkout",
  },
];
export default Header;
```

The preceding code is very similar to the code we used for the Header component in the previous chapter. It's simply a markup to display the menu items and the mini basket in the Header component.

7. Next, let us include the header in our home app.

    With Next.js, if we want code to be available within all the pages, we can create a file called _app.tsx within the /pages folder and put our relevant code in there, which is exactly what we will do to get our Header component to display across all the pages.

8. Create a new file called _app.tsx in the apps/home/pages folder with the following code:

```
import { AppProps } from "next/app";
import Head from "next/head";
import { Container } from "semantic-ui-react";
import "semantic-ui-css/semantic.min.css";
import Header from "../components/Header";

function CustomApp({ Component, pageProps }: AppProps) {
  return (
    <>
      <Head>
        <title>Welcome to ebuy!</title>
      </Head>
      <main>
        <Header />
        <Container style={{ marginTop: "5rem" }}>
          <Component {...pageProps} />
        </Container>
      </main>
    </>
  );
}

export default CustomApp;
```

9. Run pnpm dev and verify that the Header component shows up on http://
    localhost:3000.

10. Now, we will create our catalog micro app. Simply create a copy of the home app and rename
    the folder catalog.

11. Open up the catalog's package.json file, located in apps/catalog/package.json
    file, and make a few minor changes.

12. Change the app name to "name": "catalog"; let us also change the port to run on 3001:

    ```
    "dev": "next dev --port 3001".
    ```

13. Now, let us create our product card component in the components folder.

14. Create a new file in apps/catalog/components/ProductCard.tsx with the
    following code:

    ```tsx
    import { Button, Card, Image } from "semantic-ui-react";

    export function ProductCard(productData: any) {
      const { product } = productData;

      return (
        <Card>
          <Card.Content>
            <Image alt ={product.title}
      src={product.image} />

            <Card.Header>{product.title}</Card.Header>
            <Card.Description>{product.description}</Card.
    Description>
            <Card.Header>${product.price}</Card.Header>
          </Card.Content>
          <Card.Content extra>
            <div className="ui three buttons">
              <Button basic color="red">
                Remove
              </Button>
              <Button basic color="blue"></Button>
              <Button basic color="green">
                Add
              </Button>
            </div>
          </Card.Content>
    ```

```
    </Card>
  );
}
export default ProductCard;
```

Again, this is very similar to the `ProductCard` component we created in *Chapter 5*. This is a basic markup to display the product image, product name, and price, along with the add to cart button.

15. Feel free to delete the `Header.tsx` file from `catalog/components` and remove its reference from the `_app.tsx` file, as we already have it in the home app and will not be using it here.

16. Next, to save us some time, let us copy and paste the `product-list-mocks.tsx` file from *Chapter 4* into the `apps/catalog/mocks` folder. While we are here, let us also copy the `assets` folder containing the product images from `https://github.com/PacktPublishing/Building-Micro-Frontends-with-React-18/tree/main/ch4/ebuy/apps/catalog/src/assets` and paste it into `/apps/catalog/public/assets`.

17. Next, in the `apps/catalog/pages/index.tsx` file, let us add the following code:

```
import { Card } from "semantic-ui-react";
import ProductCard from "../components/ProductCard";
import { PRODUCT_LIST_MOCKS } from "../mocks/product-list-
mocks";

export function ProductList() {
  return (
    <Card.Group>
      {PRODUCT_LIST_MOCKS.map((product) => (
        <ProductCard key={product.id} product={product} />
      ))}
    </Card.Group>
  );
}

export default ProductList;
```

18. Run `pnpm dev` from the root of the `ebuy-ssr` folder and verify that the `home` and `catalog` apps work as expected. These are the URLs for our apps:

   - The **home** app: `http://localhost:3000`

   - The **catalog** app: `http://localhost:3001`

Figure 6.3 – The home micro app running on port 3000

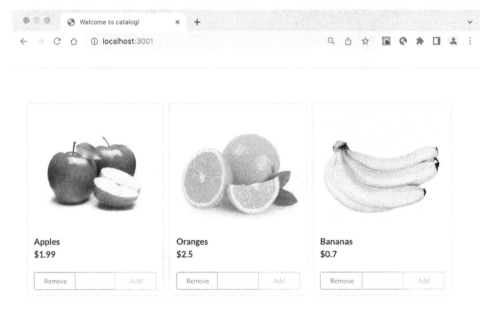

Figure 6.4 – The catalog micro app running on port 3001

Now that we have our individual apps running, let us work toward loading the catalog micro app into the home app via module federation.

Setting up Module Federation

Now that we have our apps running independently, it's time to embed the catalog app into the home app via module federation. For module federation with Next.js, we will use the dedicated nextjs-mf npm module. Follow these steps:

1. Let us first install the nextjs-mf npm module along with webpack in the catalog app:

```
pnpm add @module-federation/nextjs-mf webpack
```

2. We now need to expose the catalog app as a remote; we do this in the app/catalog/next.config.js file.

3. We replace the contents of the next.config.js file with the following:

```
const NextFederationPlugin = require("@module-federation/
nextjs-mf");
// this enables you to use import() and the webpack parser
// loading remotes on demand, not ideal for SSR
const remotes = (isServer) => {
  const location = isServer ? "ssr" : "chunks";
  return {
    catalog: `catalog@http://localhost:3001/_next/
static/${location}/remoteEntry.js`,
  };
};
module.exports = {
  webpack(config, options) {
    config.plugins.push(
      new NextFederationPlugin({
        name: "catalog",
        filename: "static/chunks/remoteEntry.js",
        exposes: {
          "./Module": "./pages/index.tsx",
        },
        remotes: remotes(options.isServer),
        shared: {},
        extraOptions: {
          automaticAsyncBoundary: true,
        },
      })
    );

    return config;
  },
};
```

Looking through the code, we first import `NextFederationPlugin`, and then we define the remote with its name and the path where its `remoteEntry.js` file can be located. Next.js creates two builds of its app – one for the server and the other for the client. Note that we conditionally load the `remoteEntry.js` file from either the `ssr` or `chunks` folder, depending on where it is executed.

4. Next, we define the webpack config where we set the properties of `NextFederationPlugin`, namely the name and what it exposes, like so:

```
exposes: {
  "./Module": "./pages/index.tsx",
  },
```

We can define an array of remotes and have different components or pages from within the catalog micro app load in other apps. This completes the setup on the catalog side.

Creating the checkout micro app

For the sake of completeness, let us also create the `checkout` micro app by creating a copy of the catalog app and renaming the folder to `checkout`. Follow these steps:

1. Let us make the necessary changes to the `apps/checkout/package.json` file, as follows:

```
"name": "checkout",
```

2. Then, update the port number:

```
"dev": "next dev --port 3002",
```

3. Now, create a file called `Basket.tsx` in `apps/checkout/components/Basket.tsx` with the following code:

```
import { Table, Image, Container } from "semantic-ui-react";

export function ShoppingBasket(basketListData: any) {
  const { basketList } = basketListData;
  return (
    <Container textAlign="center">
      <Table basic="very" rowed="true">
        <Table.Header>
          <Table.Row>
            <Table.HeaderCell>Items</Table.HeaderCell>
            <Table.HeaderCell>Amount</Table.HeaderCell>
            <Table.HeaderCell>Quantity</Table.HeaderCell>
            <Table.HeaderCell>Price</Table.HeaderCell>
          </Table.Row>
        </Table.Header>
```

```
        <Table.Body>
          {basketList.map((basketItem: any) => (
            <Table.Row key={basketItem.id}>
              <Table.Cell>
                 <Image alt={ basketItem.title } src={basketItem.
image} rounded size="mini" />
              </Table.Cell>
              <Table.Cell> {basketItem.title}</Table.Cell>
              <Table.Cell>{basketItem.quantity || 1}</Table.
Cell>

              <Table.Cell>£{basketItem.price||1 * basketItem.
quantity}</Table.Cell>
            </Table.Row>
          ))}
        </Table.Body>
      </Table>
    </Container>
  );
}

export default ShoppingBasket;
```

4. Let us also change the content of the `apps/checkout/pages/index.tsx` file to ensure that the checkout app loads the `basket` component by passing the right set of information:

```
import { Container, Header as Text } from "semantic-ui-react";
import ShoppingBasket from "../components/Basket";
import "semantic-ui-css/semantic.min.css";

import { PRODUCT_LIST_MOCKS } from "../mocks/product-list-
mocks";
export function App() {
  return (
    <Container style={{ marginTop: "5rem" }}>
      <Text size="huge">Checkout</Text>
      <ShoppingBasket basketList={PRODUCT_LIST_MOCKS} />
    </Container>
  );
}

export default App;
```

5. Now, let us update the module federation configuration in the `apps/checkout/next.config.js` to set up the checkout app as a remote.

6. Let us update the remote array to reflect the name checkout and update the port to 3002, as highlighted in the following code snippet:

```
return {
    checkout: `checkout@http://localhost:3002/_next/
static/${location}/remoteEntry.js`,
    };
The next set of changes in the same file are here
  new NextFederationPlugin({
      name: "checkout",
      filename: "static/chunks/remoteEntry.js",
      exposes: {
        "./Module": "./pages/index.tsx",
      },
  . . .
```

Let's quickly check the app to see whether the checkout app loads properly by running pnpm dev in the root folder and by visiting the following URL in the browser – http://localhost:3002.

Setting up the host app

Now, let us focus on the home app:

1. We will need to again install the module-federation/nextjs-mf npm package and webpack:

```
pnpm add @module-federation/nextjs-mf webpack
```

2. Once done, set up the host app as the host by updating the apps/home/next.config. js file, as follows:

```
const NextFederationPlugin = require("@module-federation/
nextjs-mf");
const remotes = (isServer) => {
  const location = isServer ? "ssr" : "chunks";
  return {
    catalog: `catalog@http://localhost:3001/_next/
static/${location}/remoteEntry.js`,
    checkout: `checkout@http://localhost:3002/_next/
static/${location}/remoteEntry.js`,
  };
};
module.exports = {
  webpack(config, options) {
    config.plugins.push(
      new NextFederationPlugin({
        name: "home",
        filename: "static/chunks/remoteEntry.js",
```

```
        exposes: {},
        remotes: remotes(options.isServer),
        shared: {},
        extraOptions: {
          automaticAsyncBoundary: true,
        },
      })
    );

    return config;
  },
};
```

3. Since we want to load the catalog micro app within the catalog route, we will create a new file called `catalog.tsx` in `apps/home/pages/` with the following code:

```
import dynamic from "next/dynamic";

const Catalog = dynamic(() => import("catalog/Module"), {
  ssr: true,
});

export default function catalog() {
  return <Catalog />;
}
```

4. Let us create a similar file called `checkout` in `apps/home/pages/checkout.tsx` with the following similar code:

```
import dynamic from "next/dynamic";

const Checkout = dynamic(() => import("checkout/Module"), {
  ssr: true,
});

export default function checkout() {
  return <Checkout />;
}
```

As you can see, we import Next.js's dynamic module for the first time, which is the recommended way to import dynamically with Next.js.

You can choose to dynamically import the module to execute the client side by setting up `ssr:false`; this will execute the module on the client side and be bypassed by SSR. This is suitable when your module contains personalized content, for example, recommendations, order history, and so on.

Then, we define the const called Catalog and import it from the catalog/Module. Note that the TypeScript throws an error. That's because we've not defined the types for it.

5.   So, let us quickly create the /apps/home/remotes.d.ts file with the following lines:

```
declare module "catalog/Module";
declare module "checkout/Module";
```

6.   Let's test out everything by shutting down all running servers.

killall node is a really helpful command to kill all node processes.

7.   Run pnpm dev and visit http://localhost:3000. Click on the catalog and checkout apps to see the respective micro apps load.

> **Note**
> You may need to copy the public/assets folder from the catalog into the host app.

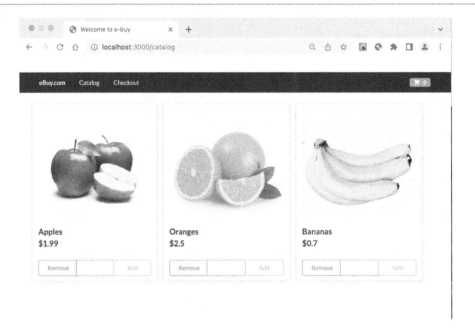

Figure 6.5 – The catalog micro app loaded in the catalog route

The following screenshot shows the checkout micro app loaded on the checkout route:

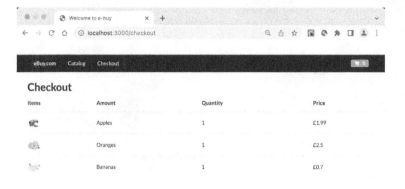

Figure 6.6 – The checkout micro app loaded on the checkout route

Congratulations!! We now have a full server side rendered microfrontend.

Let's recap what we've learned so far. We started off by creating our individual micro apps using Turborepo and Next.js, and we learned about Turborepo's folder structure and how it differs from Nx. We then created our micro apps using Next.js, and finally, we saw how to set up module federation to load the different micro apps in different routes.

Summary

We've come to the end of this chapter, where we learned about the differences between Client side rendered and server side rendered apps, and which one is suitable for which type of application. We looked at the various options to build an SSR app and zeroed in on Next.js and Turborepo to build out our module-federated app. We then saw how to set up module federation using the `next.js-mf` plugin, and we went about setting up our remote and host apps. Finally, we saw how to import these modules dynamically into the host app and set up routing between the different apps.

As a stretch goal for this chapter, you can explore setting up a shared state management solution or a shared component library, following the same approach we took in *Chapter 5*.

In the next chapter, we will learn how to go about deploying our apps to the cloud. See you on the other side!

Part 3:
Deploying Microfrontends

This part discusses strategies for deploying microfrontends, including deployment to static hosting platforms and container orchestration with Kubernetes on Azure. It covers practical considerations for deployment.

This part has the following chapters:

- *Chapter 7, Deploying Microfrontends to Static Storage*
- *Chapter 8, Deploying Microfrontends to Kubernetes*

7

Deploying Microfrontends to Static Storage

Things start to get interesting from this chapter on, because we are now stepping out of the frontend/React world and moving into the areas of cloud and full life cycle engineering.

As you may recollect from earlier chapters of this book, one of the primary goals of a microfrontend architecture is to ensure that we don't need to deploy the entire application each time a small change is made but instead only deploy the micro apps that have changed. Hence, a book on microfrontends wouldn't be deemed complete unless we covered the critical topic of deploying our microfrontend to production in the right way.

When it comes to deploying SPAs, usually we run the webpack `build` command to generate our JavaScript bundles and assets in the `/build` or `/dist` folder, which we then simply copy to a static website hosting provider to make our app available to our users. However, deploying microfrontends is a bit more complex.

In this chapter, we will see how to deploy the client-side-rendered microfrontend we built in *Chapter 5* to a static storage cloud provider such as Firebase. We will cover the following topics:

- Understanding what static storage is
- Setting up Firebase Hosting
- Learning how to build production bundles with Nx
- Learning how to only build and deploy modified apps

By the end of this chapter, we will have our microfrontend apps running on Firebase, and we will have also created scripts that only build and deploy the apps that have been modified.

Technical requirements

As we go through the code examples in this chapter, we will need the following:

- A PC, Mac, or Linux desktop/laptop with at least 8 GB of RAM (16 GB preferred)
- An Intel chipset i5+ or Mac M1+ chipset
- At least 256 GB of free hard disk storage

You will also need the following software installed on your computer:

- Node.js version 18+ (use nvm to manage different versions of Node.js if you have to)
- Terminal: iTerm2 with OhMyZsh (you will thank me later)
- IDE: We strongly recommend VS Code as we will be making use of some of the plugins that come with VS Code for an improved developer experience
- npm, Yarn, or pnpm: We recommend pnpm because it's fast and storage efficient
- Browser: Chrome/Microsoft Edge, Firefox
- A basic understanding of Nx.dev monorepos
- A basic understanding of Firebase and static site hosting would be helpful

The code files for this chapter can be found here:

`https://github.com/PacktPublishing/Building-Micro-Frontends-with-React`

We also assume you have a basic working knowledge of Git, such as branching, committing code, and raising a pull request.

What is Static Storage?

Cloud hosting providers such as AWS, Google, and Azure offer a variety of hosting solutions. Static storage, also known as blob storage, refers to a type of storage service that is optimized for storing large amounts of unstructured data, such as **Binary Large Objects** (**Blob**). This data can be of any type, including images, videos, audio files, and text file formats such as HTML, CSS, and JavaScript.

Static storage is designed to be highly scalable and is usually served via a **Content Delivery Network** (**CDN**). This allows it to handle large volumes of data without performance degradation, and also makes it highly durable, with data replication across different nodes to ensure that data is not lost due to hardware failures or other disruptions.

A key point to keep in mind about static storage is that it doesn't have any compute power; that is, it doesn't have any CPU or RAM resources. It can only serve static files. Think of it like a very large external hard disk connected to the cloud.

Historically, static storage has been used to store and serve images, JavaScript, or CSS files, or as backup storage. It was never an option to host web apps. However, with the advent of SPAs that execute on the browser, frontend engineers realized they could use storage to host JavaScript and CSS bundles and have the apps execute and run on the browser. Most hosting providers now officially offer static site hosting. Some popular static site hosting providers are the following:

- Firebase
- Netlify
- Cloudflare
- Azure Static Web Apps
- Google Cloud Storage
- Amazon S3

Due to its simplicity and very low costs, static storage is ideal for serving client-side-rendered (CSR) React apps. Due to the lack of compute power, they cannot be used to serve backend or node-based APIs, or to execute **server-side rendering (SSR)**.

In our case, since our microfrontend is client-side-rendered, we will use it to deploy our apps.

Of the various hosting options available, we will choose Firebase for our hosting solution, and in the next section, we will go about setting up our Firebase application.

Just a note that deploying the microfrontend to any other hosting provider will follow a similar process to what we will go through in the rest of the sections of this chapter.

Setting up Firebase

Firebase, which is part of Google Cloud Platform, is an extremely easy-to-use and developer-friendly hosting provider. Firebase has a lot of offerings and services for building and managing web and mobile applications.

Many of these services have free tiers, which make it ideal for building and testing things out. You can access all the products and services by heading over to `www.firebase.com` and logging in with your Google account.

Once you've logged in to Firebase, head over to **Manage Console** (`https://console.firebase.google.com/`).

Create a new project. Let's call it `ebuy`. In the next section, we will set up our sites within this project.

Setting up a project with multiple sites

We will be using Firebase's hosting service to deploy our apps. If you are not familiar with Firebase Hosting, we strongly encourage you to head over to `https://firebase.google.com/docs/hosting` and read about it:

1. Once in the console, select the **ebuy** project.

2. Head over to the **Build | Hosting** link on the left navigation pane. Click on the **Get Started** button to start the wizard and follow the steps to create a new site within the **ebuy** project.

3. We are going to need a new site for every micro app that we build, so on the Dashboard page use the **Add another site** and go ahead and create five sites. For the sake of consistency in this chapter, let's name them as follows:

    *   `ebuy-app-shell.web.app`

    *   `ebuy-catalog.web.app`

    *   `ebuy-checkout.web.app`

    *   `ebuy-recommendations.web.app`

    *   `ebuy-datastore.web.app`

Note that these names need to be unique to the entirety of Firebase. If the name is taken (and most likely it would have been taken), you can choose suitable names or go with the recommendation Firebase provides.

Once you've created these five sites, note down the URLs at which these sites will be available, as we will need them later.

Installing and configuring the Firebase CLI

Next, we need to install Firebase tools and connect them to our project and site:

1. In the terminal, run `npm install -g firebase-tools`.

2. Then, run `firebase login`. This will open up a browser window and request you to log in to your Firebase account.

3. Run `firebase init hosting`. This will take you through a series of steps. If all goes well, then you will see new `.firebaserc` and `firebase.json` files created.

4. Next, we need to let Firebase know which micro app should be deployed to which target site. We do this by running the following commands. The syntax looks as follows:

    ```
    firebase target:apply hosting <micro-app-name> <firebase-
    site-name
    ```

5. So, in our case, given the names we have for our micro apps and the websites created within Firebase, our commands would look as follows:

    A. `firebase target:apply hosting app-shell ebuy-app-shell`

    B. `firebase target:apply hosting catalog ebuy-catalog`

    C. `firebase target:apply hosting checkout ebuy-checkout`

    D. `firebase target:apply hosting recommendations ebuy-recommendations`

    E. `firebase target:apply hosting store ebuy-datastore`

Once these commands have been executed successfully, you'll notice these entries being made in the `.firebaserc` file.

This completes our setup on the Firebase side of things. In the next section, we will prepare our microfrontend for production builds.

Creating the Microfrontend Production build

As you may recollect, so far, we've only run and tested our microfrontends in development mode, using the `nx serve` command. For us to deploy applications to a hosting server, they need to be built in production mode.

This is usually quite straightforward in regular React apps, but with our microfrontends, it needs a bit more work.

Open up the ebuy app we built in *Chapter 5* and follow these steps. Let's first create a script command to build all our apps:

1. Open up the `package.json` file on the root and just like the `serve:all` command, let's create a new command for `build:all` as follows:

    ```
    "build:all": "nx run-many --target=build"
    ```

2. Run the `pnpm build:all` command and let us see whether all the apps build. Oops! You'll notice while all the other apps built fine, `app-shell` threw out some error about not being able to find `catalog/Module` or `checkout/Module`, and so on.

    Let's dig a bit into it.

3. Open up the `/apps/app-shell/project.json` file and have a look at the build scripts object. You will notice that it uses a different `webpackConfig` file for production builds, namely the one located here: `apps/app-shell/webpack.config.prod.js`.

Let's open up this file and have a look. In there, you will notice that the `remotes` array is blank. This is the reason why our app-shell build command is failing, because webpack doesn't know the path from where it needs to fetch the `remoteEntry.js` file.

4.   Let's add our list of remotes to this array. This should mirror the list of apps in the remotes array of our `module-federation.config` file.

When entering these remote paths, since module-federation is now going to pick them from a publicly hosted URL, we will need to use the full path for where these `remoteEntry` files will exist.

5.   We update the remotes array in `apps/app-shell/webpack.config.prod.js` as follows:

```
remotes: [
  ['catalog', 'https://ebuy-catalog.web.app/'],
  ['checkout', 'https://ebuy-checkout.web.app/'],
  ['store', 'https://ebuy-datastore.web.app/'],
]
```

6.   Now, rerun the `pnpm build:all` command and verify that all the apps build successfully.

Our work isn't fully done yet. As you will recollect, our `catalog` and `recommendations` apps also need the array of remotes in their `webpack.config.prod.js` files.

We also notice that because our catalog and checkout apps were not originally built as microfrontend remote apps, they have a slightly different configuration and are missing `webpack.config.prod.js` files. Let's fix that first.

7.   First and foremost, let's copy and paste the `webpack.config.prod.js` files from the `app-shell` app into our catalog, checkout and recommendations apps.

Then, we need to let the builder know that we want webpack to pick up the configurations from this `.prod.js` file when building the production builds.

8.   So, in their respective `project.json` files, we add the following line within the `build > configuration > production` object, as follows:

```
//apps/catalog/project.json

  . . .

"vendorChunk": false,
  "webpackConfig": "apps/catalog/webpack.config.prod.js"
        },


//apps/checkout/project.json
```

```
. . .
    "vendorChunk": false,
      "webpackConfig": "apps/checkout/webpack.config.prod.js"
          },
```

This will now ensure that all apps use their corresponding `webpack.config.prod.js` file to run their production builds.

9. Now, let's go and update the array for the remote paths in our `apps/catalog/webpack.config.prod.js` file. Since the catalog app has only one remote, which is the recommendations micro app, our remotes array would look like this:

```
remotes: [
   ['recommendations', 'https://ebuy-recommendations.
web.app/'],
 ],
```

Next, let's do the same for our recommendations apps, which use the `store` micro app as a remote. So, in the `apps/recommendations/webpack.config.prod.js` file, we update the remotes array as follows:

```
remotes: [
   ['store', 'https://ebuy-datastore.web.app/'],
 ],
```

10. 10. Since the checkout app also needs to use the store as a remote we update the `apps/checkout/webpack.config.prod.js` as follows:

```
remotes: [
   ['store', 'https://ebuy-datastore.web.app/'],
 ],
```

11. Run our `pnpm build:all` command again to generate the production builds based on the latest webpack configurations we made.

When the build is successful, have a look in the `/dist` folder on the root of the project and verify that all our micro-app folders are present within `/dist/apps`. Note their paths as we will need them in the next section.

In this section, we were able to generate production builds for our microfrontends by ensuring all the apps used the right webpack configuration, including the correct public URLs for the `remoteEntry.js` files.

In the next section, we will see how to deploy these apps to Firebase.

Deploying our Apps to Firebase

Deploying our apps to Firebase is quite easy using the Firebase CLI's `deploy` command. However, before we run our Firebase `deploy` command, we need to let Firebase know which micro-apps go into the corresponding Firebase website. We do this in the `/firebase.json` file.

Replace the default configuration with the following:

```json
{
  "hosting": [
    {
      "target": "app-shell",
      "public": "dist/apps/app-shell",
      "ignore": ["firebase.json", "**/.*", "**/node_
modules/**"],
      "rewrites": [
        {
          "source": "**",
          "destination": "/index.html"
        }
      ]
    }
    {
      "target": "catalog",
      "public": "dist/apps/catalog",
      "ignore": ["firebase.json", "**/.*", "**/node_
modules/**"],
      "rewrites": [
        {
          "source": "**",
          "destination": "/index.html"
        }
      ]
    },
    {
      "target": "checkout",
```

```
      "public": "dist/apps/checkout",
      "ignore": ["firebase.json", "**/.*", "**/node_
modules/**"],
      "rewrites": [
        {
          "source": "**",
          "destination": "/index.html"
        }
      ]
    },
    {
      "target": "recommendations",
      "public": "dist/apps/recommendations",
      "ignore": ["firebase.json", "**/.*", "**/node_
modules/**"],
      "rewrites": [
        {
          "source": "**",
          "destination": "/index.html"
        }
      ],
    },
    {
      "target": "store",
      "public": "dist/apps/store",
      "ignore": ["firebase.json", "**/.*", "**/node_
modules/**"],
      "rewrites": [
        {
          "source": "**",
          "destination": "/index.html"
        }
      ]
    }
  ]
}
```

As you can see, the preceding code is a configuration, where we have an array of our target apps, and we define the folder path where Firebase should look for the bundles for each of the micro apps. We also have a few settings regarding ignoring and not deploying node_modules and a rewrite rule, which is essential if you want each micro app to also be available as its own SPA within its respective site.

With this, we are now ready to deploy our apps to Firebase. Let's first run it manually to ensure things work fine.

In the terminal of the project, run the following command:

```
firebase deploy --only hosting
```

This would generate a .firebase folder with a lot of files. Don't forget to add .firebase to your .gitignore.

Let Firebase do its thing, and if all goes well, it should display a success message and print out the list of URLs where the sites have been deployed, like so:

```
✓  Deploy complete!

Project Console: https://console.firebase.google.com/project/ebuy-90233/overview
Hosting URL: https://ebuy-appshell.web.app
Hosting URL: https://ebuy-catalog1.web.app
Hosting URL: https://ebuy-checkout1.web.app
Hosting URL: https://ebuy-recommendations1.web.app
Hosting URL: https://ebuy-store1.web.app
```

Figure 7.1 – List of website URLs published after a successful deployment on Firebase

Great! Let's click on the app-shell link and check whether we can see our microfrontend.

Err, we see a blank page... Have a peek into the browser's developer tools console and you'll notice what the problem is. Our browser has blocked calls to the remoteEntry.js files because of **Cross-Origin Resource Sharing** (**CORS**).

Figure 7.2 – CORS policy headers due to missing Access-Control-Allow-Origin header

We will see how to fix this in the next section.

Fixing CORS issues

If you've been building React or any other web apps, you'll be familiar with the dreaded CORS problem. This is where the browser, for security reasons, prevents calls to external domains unless it sees an explicit `'Access-Control-Allow-Origin'` header. The access control is set on the apps that decide whether they want assets from their domain to be consumed and executed on other domains.

So, for our microfrontend apps to work properly, the host app needs to be able to load the `remoteEntry.js` file from the public URL where each of the micro apps is hosted. This is what we are going to set in the next steps.

With Firebase Hosting, it is quite easy, and we can define a headers array in the `firebase.json` file.

Open up `/firebase.json` for all the apps except `app-shell`, and within each of the target objects, define the headers we'd like to set:

```
    "headers": [
      {
        "source": "**/*.@
(eot|otf|ttf|ttc|woff||woff2|js|font.css|remoteEntry.js)",
        "headers": [
          {
```

```
            "key": "Access-Control-Allow-Origin",
            "value": "https://ebuy-app-shell.web.app"
         }
       ]
     }
   ]
```

What we are basically saying here is each of the micro apps allows the list of defined file types to be called and executed from `https://ebuy-app-shell.web.app`.

Note that we need to add the headers array for every target app defined within the `firebase.json` file.

Rerun `firebase deploy --only hosting` and now, you should be able to view all the sites working on `https://ebuy-app-shell.web-app/`.

Deploying only the selected target

Currently, the `firebase` command deploys all the micro-apps. If we wanted to deploy only one of the micro apps, we'd simply need to pass the target name as an argument:

```
firebase deploy --only hosting:<app-name>
```

So, if we wanted to deploy only `app-shell`, our command would look as follows:

```
firebase deploy --only hosting:app-shell
```

This will be critical in the next section.

Looking back at this section, we were able to deploy our apps to Firebase, and we also managed to fix the CORSs issue by setting `Access-Control-Allow-Origin` headers. We also saw the CLI syntaxes that allow us to deploy only the apps that we need.

In the next section, we will use these CLI commands in combination with another nifty command from Nx to control and deploy only the apps that changed.

Deploying only Micro Apps that changed

To be able to deploy only the micro apps that have been impacted by modifications to a file, we basically need to be able to do two things:

1. Identify which apps have been impacted due to changes to a given set of files
2. Only build and deploy the micro-apps that have been impacted

For the second point, from the previous section, we now know how to let the Firebase CLI know which micro-app we would like to be deployed. We will look at how to achieve the first point in the next subsection.

NX Affected

The NX dev tools come with a handy command called `nx affected`, which is able to keep track of what files changed from the previous commit and highlight the apps that have been impacted due to the changes to these files.

This is a nifty feature that can be used for various purposes, such as speeding up the execution of tests by running unit tests or build commands only against projects that have been impacted by changes to certain files – or, in our case, deploying only the micro-apps that have changed.

To give it a quick try, run `git add.` and `git commit` to commit all the changes we have made so far. Try and make a small visual change to `apps/app-shell/src/app/app.tsx`. Save the file and run the following command:

```
pnpm nx print-affected --type=app --select=projects
```

It should print out `app-shell` as the app that was modified. Now, try and make changes to `libs/mocks/src/lib/product-list-mocks.tsx` and run the same command. You will see the catalog and checkout apps also added to the list of apps that are affected.

The way the `nx affected` command works is by comparing the difference between the SHAs of the main branch and the current HEAD. You can pass in additional parameters to the affected command to compare the difference between any base and head and run a command passed to the target flag:

```
pnpm nx affected --target=deploy --base=main --head=HEAD
```

`--target` is the custom command to run, `--base` is the base you want to compare against, and `--head` is the tip of your Git branch.

This will probably return a message saying **Nx successfully ran target deploy on 0 projects**. This is because we haven't created our custom deploy command yet.

To get a deeper understanding of the various options for `nx affected`, have a read here: `https://nx.dev/nx/affected#affected`.

In addition to affected, you may also find the `nx graph` command useful for getting a nice, visual representation of the various micro-apps consuming the different shared components and utilities form the `libs` folder.

Try running `pnpm nx affected:dep-graph` to get a visual graph of how the modified files impact the micro-apps.

Here is an example of how changes to the `libs/mocks/src/lib/product-list-mocks.tsx` file impact both the catalog and checkout apps, because both these apps import the product list from the `product-list-mocks` file:

Figure 7.3 – nx affected:dep-graph highlighting the projects impacted due to a change in mocks

> **Note**
>
> nx graph or nx affected doesn't take into account the host and remote features of module federation.

Creating an Nx custom command executor to deploy

Executors in Nx allow you to create custom script commands for a project, which you can run via the Nx command system.

Please do take the time to read more about Nx custom command executors here: `https://nx.dev/recipes/executors/run-commands-executor#3.-run-the-command`.

Let's create a custom command to deploy an individual micro app.

In `apps/app-shell/project.json`, add the following code within the target attribute:

```
"deploy": {
  "executor": "@nrwl/workspace:run-commands",
  "options": {
    "commands": ["firebase deploy --only hosting:app-
```

```
shell"],
      "parallel": true
   }
}
```

Add the deploy custom command to each of the micro-app's `project.json` files. Pass the correct micro-app name in the argument.

Once that is done, try making a small change in the mocks file and run the following two commands:

```
pnpm nx affected --target=build
pnpm nx affected --target=deploy
```

Assuming Nx has detected the difference correctly, it will only build the catalog and checkout apps and you will also notice that these are the only two apps that deployed to Firebase.

You can verify that by going into Firebase Console's hosting dashboard and checking the timestamp of when the apps were last deployed:

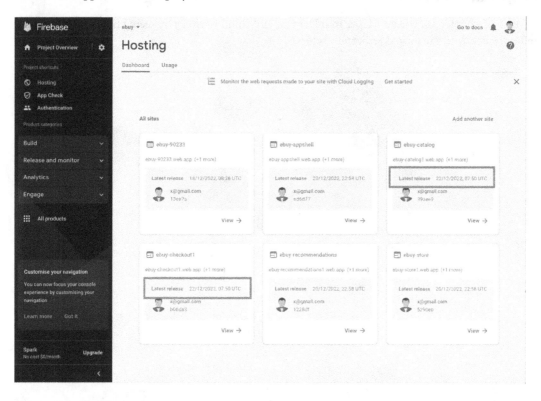

Figure 7.4 – Firebase Console displaying the deployed timestamp of modified apps

Navigate to `https://ebuy-app-shell.web-app/` (use the correct URL as displayed in your Firebase Console) and verify that everything continues to work fine and that the changes you've made reflect on the app. You may need to do a hard reload on your browser to view the updates.

And with this, we've successfully managed to deploy only the apps that have changed while ensuring that the rest of the app works as expected.

Summary

With that, we come to the end of this chapter, where we learned about static storage hosting and why it is ideal for deploying and serving client-side-rendered React apps. We saw how to build production bundles for our module-federated micro app. We then saw how to set up a multi-site project in Firebase and used Firebase CLI commands to deploy our apps. We also saw how to address CORS issues by setting the right header values for the `Access-Control-Allow-Origin` header, and then finally, we saw how to combine the `nx affected` command and Firebase's `hosting:<app-name>` command to detect the micro-apps that have been impacted by a change and only build and deploy them to Firebase. We also used this as an opportunity to create a custom command executor to deploy these affected apps.

In the next chapter, we will go deeper into DevOps and cloud territory by seeing how to deploy our microfrontends to a managed Kubernetes cluster.

8

Deploying Microfrontends to Kubernetes

In the previous chapter, we learned how to manually deploy our microfrontends to a static storage provider such as Firebase.

In this chapter, we will go deeper into cloud and DevOps territory by learning how to deploy our apps to a managed Kubernetes cluster. Kubernetes has become the de facto choice to deploy enterprise-grade web apps (both backend and frontend) to the cloud.

When it comes to deploying SPAs, we run usually the webpack `build` command to generate our JavaScript bundles and assets in the `/build` or `/dist` folder, which we then simply copy to a static website hosting provider to make our app available to our users. However, deploying microfrontends is a bit more complex.

In this chapter, we will see how to deploy our module-federated microfrontend to a managed Kubernetes cluster.

We will cover the following topics:

- How to containerize our apps using Docker
- The basics of Kubernetes and its various components
- Some basic commands to manage our Kubernetes cluster
- DevOps and how to automate deploying our micro-apps to Kubernetes

By the end of this chapter, we will have our microfrontend apps running on a Kubernetes cluster in Azure. We will deploy them via an automated **Continuous Integration** (**CI**) and **Continuous Delivery** (**CD**) pipeline that will automatically build and deploy the necessary apps whenever code is merged.

Technical requirements

In addition to all the standard technical requirements that we mentioned in the previous chapters, you will need the following:

- An Azure cloud subscription

- Access to GitHub and GitHub Actions

- A high-level understanding of CI and CD concepts

- Knowledge of Docker and containerizing apps will be helpful

The code files for this chapter can be found at the following URL, where we essentially started with the microfrontend we built in *Chapter 6*: `https://github.com/PacktPublishing/Building-Micro-Frontends-with-React`.

We also assume you have a basic working knowledge of Git, such as branching committing code and raising a pull request.

Introduction to Kubernetes

Kubernetes, also known as **K8s**, has taken the cloud and DevOps world by storm. Originally developed by Google and now part of the Cloud Native Computing Foundation, Kubernetes provides all the tools necessary to deploy and manage large-scale, mission-critical applications on the cloud from a single interface.

Traditionally, managing a large-scale, production-grade application on the cloud meant having to deal with things such as web servers, load balancers, auto-scaling, and internal and external traffic routing. Kubernetes now brings all of that under a single umbrella and provides a consistent way to manage all the components of a cloud environment.

The premise of Kubernetes is that you tell it the end state of what you want via a spec file, and Kubernetes will go about getting it done for you. For example, if you tell Kubernetes that you want three replicas for your application with a service load balancer, Kubernetes will figure out how to spin up the three replicas and ensure that the traffic is equally distributed between the three replicas. If, for some reason, one of the pods restarts or shuts down, Kubernetes will automatically spin up a new pod to ensure that, at any given time, three replicas of the pod service traffic. Similarly, when you deploy a new version of the app, Kubernetes will take over the responsibility of gradually spinning up new pods with the latest version of the app, while gracefully shutting down the pods with the older version of the application.

Through the rest of this section, we will look at some of the key components of Kubernetes that apply to us, along with the architecture to deploy our microfrontend on Kubernetes.

What is Kubernetes?

Kubernetes is a platform-agnostic container orchestration platform that enables the deployment, scaling, and management of containerized applications in a cluster of machines.

It abstracts the underlying infrastructure, allowing you to run your applications in a variety of environments, including on-premises data centers, public cloud providers such as Microsoft Azure, Google Cloud Platform, and Amazon Web Services, and even on your own laptop.

Kubernetes is designed to be highly modular and extensible, and it integrates with a variety of tools and services to support the complete life cycle of an application, including deployment, scaling, monitoring, and maintenance. It is widely adopted in the industry and has become the de facto standard for container orchestration.

Key concepts of Kubernetes

Kubernetes can be quite a vast topic and would need a dedicated area of focus to go deep into it. You can go into the details of the various components of Kubernetes here: `https://kubernetes.io/docs/concepts/overview/components`. However, as a frontend engineer and for the scope of this book, there are six basic concepts and terms that you need to be aware of:

- **Nodes**: A node is a worker machine in a Kubernetes cluster. It can be a physical or virtual machine, and it is responsible for running the containerized applications deployed to it.

- **Pods**: A pod is the basic execution unit of a Kubernetes application. It is a logical host for one or more containers, as well as all containers in a pod run on the same node. Pods provide a shared context for containers, such as shared storage and networking.

- **Services**: A service is a logical abstraction over a group of pods. It defines a policy to access the pods, typically via a stable IP address or DNS name. Services allow you to decouple the dependencies between your applications, enabling you to scale or update a group of pods without affecting the consumers of the service.

- **Deployments**: A deployment is a declarative way to manage a ReplicaSet, which is a set of identical pods that are deployed to the cluster. Deployments allow you to specify the desired state of your application, and Kubernetes will ensure that the actual state matches the desired state. This includes rolling updates, rollbacks, and self-healing.

- **Ingress**: Ingress is a way to expose your services to the external world. It provides a way to map external traffic to a specific service in your cluster, typically via a stable IP address or DNS name. Ingress can also provide additional features, such as SSL termination and load balancing. Think of it as a router where a URL is mapped to a service.

- **Namespaces**: A namespace is a logical partition in a Kubernetes cluster. It allows you to use the same resources (such as names) in different contexts, and it can be used to isolate resources within a cluster.

Kubernetes architecture for microfrontends

When deploying our microfrontends on Kubernetes, we create a pod for each micro app, and this micro app is exposed internally via an Ingress service.

The home app module federates all these micro-apps. The following diagram helps to explain the architecture better:

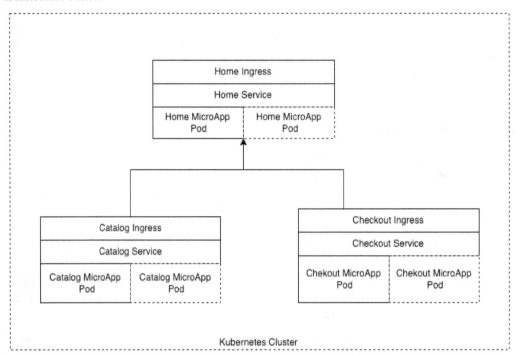

Figure 8.1 – Kubernetes topology architecture to deploy microfrontends

As you can see in *Figure 8.1*, each of our micro-apps is deployed within its own pod. These pods can be replicated or set to auto-scale as traffic increases. This is denoted by the dotted box around the pod. These pods are exposed via a service, which acts as a sort of load balancer. Therefore, the home app service is the single endpoint for all the replications of the home micro app pod.

Each of the services is exposed via an Ingress route. This is where we define the URL for our micro app, which eventually will be used in our module federation configuration. This is what the overall Kubernetes architecture will look like.

With this, we come to the end of this section, where we learned about some of the key concepts of Kubernetes, such as nodes, pods, services, Ingress, and the architecture of our micro-apps within a Kubernetes cluster. In the next section, we will see how to go about containerizing our app so that it can be deployed into a Kubernetes cluster.

Containerizing our micro-apps with Docker

Containers are a way to package software applications in a standardized and portable way, allowing them to run consistently across different environments. They provide a lightweight and efficient way to run applications and are particularly well-suited for microservices architectures, where an application is composed of multiple, independently deployable services.

In this section, we will look at how to install Docker and create a Docker image by creating a Dockerfile.

Installing Docker

Docker Engine is available for personal use on multiple Linux, Mac, and Windows systems via Docker Desktop. You can follow the instructions here to install the Docker engine: https://docs.docker.com/engine/install/.

> **Note**
>
> If you don't want to or can't use Docker Desktop on your Windows or Mac, there are alternatives, such as Rancher Desktop, Podman, and Colima.

Once you have Docker installed, verify it by running the following command in the terminal:

```
docker -v
```

If it returns the version of Docker, then you are all set, and it means that Docker was installed successfully on your system.

Creating standalone app builds

Before we can start creating a Docker image, we will first need to ensure that the build outputs of our micro apps are self-contained and can run in standalone mode. We do this by adding the following lines in each of the next.config.js files, like so:

```
const path = require("path");

module.exports = {
  output: "standalone",
  experimental: {
    outputFileTracingRoot: path.join(__dirname, "../../"),
  },
```

```
...
}
```

`outputFileTracingRoot` is an experimental feature introduced in Next.js 12+ onward; this helps reduce the size of the build outputs, especially when we want to try and reduce our Docker image sizes.

Make sure to add these lines to the `next.config.js` file for each of the micro apps.

Creating a Dockerfile

The next step is to create our Dockerfile, which contains the instructions for Docker to create our Docker image.

Since we need to create a Docker image for each micro app, we will create a Dockerfile within `apps/home`. The default filename we usually give to this is `Dockerfile`.

Let's add the following commands to this Dockerfile. We will use the default Dockerfile provided by Turborepo and Next.js.

We will build our Dockerfile as a multi-stage file, which allows us to leverage the caching of the layers and also ensures that the size of the Docker image is as small as possible.

We will build it in three stages, starting with the builder stage:

```
FROM node:18-alpine AS base

FROM base AS builder
# Check https://github.com/nodejs/docker-node/tree/
b4117f9333da4138b03a546ec926ef50a31506c3#nodealpine to
understand why libc6-compat might be needed.
RUN apk add --no-cache libc6-compat
RUN apk update
# Set working directory
WORKDIR /app
RUN yarn global add turbo
COPY . .
RUN turbo prune --scope=home --docker
```

As you can see, we use a base image of Node Alpine 18.14, and we call it the builder stage. Alpine is the most minimalistic version of Node.js.

Now, we install the `libc6-compact` library and run the `update` command. Then, we set the working directory for the app and install turbo.

Then, we copy everything from our repo (note the space between the two periods in the COPY command).

Finally, we run the turbo prune command to extract all the files necessary for the home micro app.

Now, we will move on to the installer stage and continue writing the following code immediately after the previous code:

```
FROM base AS installer
RUN apk add --no-cache libc6-compat
RUN apk update
WORKDIR /app
# First install the dependencies (as they change less often)
COPY .gitignore .gitignore
COPY --from=builder /app/out/json/ .
COPY --from=builder /app/out/pnpm-lock.yaml ./pnpm-lock.yaml
RUN yarn global add pnpm
RUN pnpm install --no-frozen-lockfile
# Build the project
COPY --from=builder /app/out/full/ ./
COPY turbo.json turbo.json
RUN ENV=PROD yarn turbo run build --filter=home...
```

Again, we start by defining the base image as the installer, running the regular apk add and update commands, and setting the working directory.

Then, we copy the .gitignore file as well as the relevant files from the /app/out folder from the builder stage.

We then install pnpm and run the pnpm install command.

Then, we copy all the files from the app/out/full folder from our builder stage and run the turbo build command.

Then, we move on to the final runner stage where we write the following code:

```
FROM base AS runner
WORKDIR /app

# Don't run production as root
RUN addgroup --system --gid 1001 nodejs
RUN adduser --system --uid 1001 nextjs
```

```
USER nextjs

COPY --from=installer /app/apps/home/next.config.js .
COPY --from=installer /app/apps/home/package.json .


# Automatically leverage output traces to reduce image size
# https://nextjs.org/docs/advanced-features/output-file-tracing
COPY --from=installer --chown=nextjs:nodejs /app/apps/home/.
next/standalone ./
COPY --from=installer --chown=nextjs:nodejs /app/apps/home/.
next/static ./apps/home/.next/static
COPY --from=installer --chown=nextjs:nodejs /app/apps/home/
public ./apps/home/public


CMD node apps/home/server.js
```

In the preceding code, we basically create a user group to avoid the security risks of running the code as root, and then we copy the relevant files from our installer stage and run the `node` command.

Now, we need to create a `.dockerignore` file in the root of the repo, where we list the files and folders that we don't want Docker to copy to the image:

```
node_modules
npm-debug.log
**/node_modules
.next
**/.next
```

Let's test the Dockerfile to see whether it builds. From the root of the application, run the following command in the terminal:

```
docker build -t home -f apps/home/Dockerfile .
```

`-t` stands for the tag name, and it will create a Docker image with the name home. The `-f` part is the path to the Dockerfile.

Note the space and period at the end of the command, which is important. The period at the end denotes the build context – that is, the set of files and folders Docker should use to build the image. The period also denotes that we want to package all the files and folders in the current directory.

This command will take several minutes to run on its first time, as Docker will download the base node image and other dependencies. The subsequent builds will be a lot faster, as Docker will cache the layers and reuse them if the layer hasn't changed.

You can run the Docker image locally by running the following command:

```
docker run -p 3000:3000 home
```

Once we've verified that this works fine, we will need to create similar Dockerfiles for each of our apps.

So, in `apps/catalog` and `apps/checkout`, copy and paste the Dockerfile and replace all instances of `home` with the relevant micro app name.

Note that each of these micro apps runs on the same port, `3000`, so to test them locally, we can test only one image at a time, unless you change the `hostPort` value to something different or use a docker-compose file.

Now that we have learned how to dockerize our micro apps and run them locally, we will move on to the next section on setting up Docker Hub.

Setting up Docker Hub to store Docker images

In the previous section, we created Docker images of our apps and were able to run them locally. For us to be able to deploy them on Kubernetes, we need to store them in a container library from where our DevOps pipelines can pull the images. We will use a free artifact registry solution such as Docker Hub for this. Alternatively, you can use other container registry solutions provided by various hosting providers, such as Azure Container Registry, Google Container Registry, and Amazon Elastic Container Registry:

1. Log in/register at `https://hub.docker.com`, and then create three public repositories one for each micro-app. We will call them the following:

    * `ebuy-home`

    * `ebuy-catalog`

    * `ebuy-checkout`

2. Make a note of the Docker registry paths, which are usually of the `<your-username>/ebuy-home` format, `<your-username>/ebuy-catalog` format, and so on.

3. Then, let's create an access token that will be needed for our CI and CD pipelines. Go to **Account Settings**, and on the **Security** page, create a new access token and give it a description. Under **Access permissions**, select **Read and Write**, as our pipelines will need to push and pull the Docker images.

4. Once the token is generated, copy and keep it safe, as it will never be displayed again. (You can always generate a new token if you've lost the old one.)

Our work on Docker Hub is done!

In the next section, we will create our Kubernetes configuration files that will be used to spin up our Kubernetes cluster.

Creating a Kubernetes configuration file

Earlier in this chapter, in the *Introduction to Kubernetes* section, we learned about the various Kubernetes services that we will use to deploy our microfrontends.

Deploying these services on Kubernetes is commonly done by defining the various configuration settings in a `.yaml` file and then applying the configuration to the Kubernetes cluster.

In this section, we will learn about the structure of these Kubernetes spec files and how to go about creating them for our deployments, services, and Ingress.

The structure of a Kubernetes spec file

A Kubernetes spec file is a YAML document that describes the desired state of a Kubernetes object, such as a Deployment, Pod, Service, or ConfigMap. The structure of a Kubernetes spec file generally consists of two main parts – the metadata section and the spec section. Each file always starts by defining the `apiVersion` and the `kind` of spec file.

The metadata section includes information about the object, such as its name, labels, and annotations. This section is used by Kubernetes to manage the object and enable other objects to reference it.

The spec section includes the desired state of the object, such as the container image, resource requests and limits, networking configuration, and any other relevant settings. This section is used by Kubernetes to create and manage the object according to its desired state.

Creating spec files to deploy our microfrontends

As we saw earlier, the structure of a Kubernetes spec file follows a hierarchical format, with each section and its corresponding properties nested under the appropriate heading. Additionally, many Kubernetes objects have properties that are specific to their type, so the structure of the spec file may vary depending on the object being described.

Let's start by creating these files in a folder called `k8s` within each of the micro apps folders.

Let's start by creating the `/apps/home/k8s/deployment.yml` file with the following code. The `deployment.yml` file contains the configuration to set up and configure the Kubernetes pods within which our micro app will run:

```
apiVersion: apps/v1
kind: Deployment
metadata:
  name: home
  namespace: default
  labels:
    app: home
spec:
  replicas: 1
  selector:
    matchLabels:
      app: home
  template:
    metadata:
      labels:
        app: home
    spec:
      containers:
        - name: home
          image: <dockerUserID>/ebuy-home:latest
          imagePullPolicy: Always
          ports:
            - name: http
              containerPort: 3000
              protocol: TCP
```

As you read through the `deployment.yml` configuration file, you will see that we label the app as home and also use the same name to define the name of our container. We define the number of replicas as one, which means it will spin up one pod; increase this number to two or more if you want multiple replicas of the pod. Then, within the container section of the file, we define the name of the path of the Docker image it should use and the ports and protocols that it should use. Replace this with the values of your Docker repository. Note `:latest` at the end of the Docker image value; this is something we add to ensure that Kubernetes always picks up the latest version of the Docker image.

Now, we define the service, which acts as a sort of load balancer over one or more replicas of the pod.

Create a new file called `/apps/home/k8s/service.yml` with the following code:

```
kind: Service
apiVersion: v1
metadata:
  name: home
  namespace: default
  labels:
    app: home
spec:
  type: LoadBalancer
  selector:
    app: home
  ports:
    - protocol: TCP
      port: 80
      targetPort: 3000
      name: home
```

The `service.yml` file is quite straightforward, wherein we provide the necessary metadata such as the `name`, `label`, and `namespace` for the Kubernetes cluster.

Then, within the specs, we define what type of service this is; we will set it as a `LoadBalancer`. This will help expose a public IP address that we will need later and, finally, within the `ports` section, the protocol and port numbers on which we will expose the service.

Finally, we need to define the `ingress.yml` file where we will assign a URL to the service. Create a file called `/apps/home/k8s/ingress.yml` with the following code.

The Ingress within Kubernetes essentially runs nginx under the hood, so if you are familiar with nginx, configuring this should be easy:

```
apiVersion: networking.k8s.io/v1
kind: Ingress
metadata:
  name: home
  namespace: default
```

```
    labels:
      app: home
    annotations:
      # nginx.ingress.kubernetes.io/enable-cors: 'true'
      # nginx.ingress.kubernetes.io/cors-allow-origin: '*'
      nginx.ingress.kubernetes.io/rewrite-target: /$2
  spec:
    ingressClassName: nginx
    rules:
    - http:
        paths:
        - path: /home(/|$)(.*)
          pathType: Prefix
          backend:
            service:
              name: home
              port:
                number: 80
```

This is generally a bit of a tricky file to configure, as this is where you define the URL structures and rewrite rules and other nginx configurations as you'd do for a web server. As you can see, we define the regular metadata information under annotations, and we define the various rewrite rules and nginx configurations, such as CORS. Then, we set the `regex` path, which tells Kubernetes through which URLs it should direct traffic to this service and pod. Finally we need to copy and paste the K8s folder into each of our micro apps and update the relevant paths and app names to match the name of the micro app.

As we come to the end of this section, we've seen how to create Kubernetes spec files to deploy pods, how to set up a service that sits over these pods, and finally, the ingress that provides routing to these pods. In the next section, we will create an Azure Kubernetes cluster, against which we will execute these specs.

Setting up a managed Kubernetes Cluster on Azure

In this section, we will learn how to set up a managed Kubernetes cluster on Azure. The reason it's called *managed* is because the master node, which is sort of the brain of Kubernetes, is managed by Azure, and we only need to spin up the worker nodes. We will see how to log in to Azure and create a subscription key, and we will install Azure CLI and collect the various credentials that we need for our DevOps pipeline.

For this chapter, we will use **Azure Kubernetes Service** (**AKS**) to set up our cloud-based managed Kubernetes cluster. You can also set up a managed Kubernetes cluster on Google Cloud using **Google Kubernetes Engine** (**GKE**), or you can use Amazon **Elastic Kubernetes Service** (**EKS**) on AWS.

Irrespective of whichever hosting provider you use to set up your Kubernetes cluster, the Dockerfile and the Kubernetes configuration `.yaml` files remain the same.

Logging into the Azure portal and setting up a subscription key

To carry out any activity on the Azure platform, you need to have the login credentials for the platform and a subscription key. All the resources that we create within Azure need to be mapped to a subscription key, which eventually is used by Azure to calculate the hosting charges. To do this, follow these steps:

1. Head over to `https://portal.azure.com` and log in with a Microsoft login; if you don't have one, you can always sign up for one.

2. Once logged into the portal, search for `Subscriptions` and add a **Pay-As-You-Go** subscription. If you have an **Azure for Student** or free trial subscription in your list, feel free to select either one of them as well. This subscription will be used for all the hosting costs that will be incurred as part of the various services you run within Azure.

3. Then, in the search box, search for `Resource Group` and create a resource group. Let's call it `ebuy-rg`; the `rg` suffix stands for **resource group**. It would have selected the default subscription that you created in the earlier step. For the region, you can select **US East** or a region of your choice; for the sake of consistency in this chapter, we will stick with **US East**.

    In Azure, it is always a good practice to create a resource group for a project and then have all the various services associated with that project within the resource group. This allows us to easily manage the resources within the resource group, especially when we want to shut down all the services for the project.

4. Next, we will create our AKS cluster; search for `Azure Kubernetes Service (AKS)`, click on the **Create** button in the top-left corner, and then select the **Create a Kubernetes cluster** menu item. You will be presented with a screen, as shown in the following screenshot:

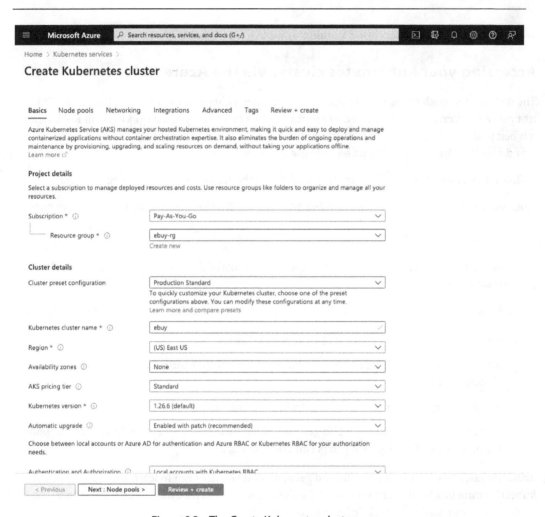

Figure 8.2 – The Create Kubernetes cluster screen

5. Select the subscription and resource group we created in the earlier steps, and then, in the Cluster preset configuration, select **Production Standard** as the preset configuration. You can also choose other higher configurations; however, note that the AKS cluster is the most expensive component of your Azure monthly billing.

6. Provide the Kubernetes cluster name as ebuy, and select the same region where you have your resource group created; in our case, it is **(US) East US**. For the Kubernetes version, you can choose to leave it as default or select **1.26.6** to ensure the settings are consistent with the code and configuration defined in the chapter. For the scale method, set it to **Autoscale**, and for the maximum number of nodes, leave it at **1** or **2**. Finally, hit **Review + Create**, and then after the validation check is done, hit **Create**.

We now have our Kubernetes cluster running within AKS.

Accessing your Kubernetes cluster via the Azure CLI

The de facto approach to interacting with your Kubernetes cluster on Azure is via the Azure CLI at `https://learn.microsoft.com/en-us/cli/azure/`. If you are working with Kubernetes it is best to also install kubectl, the instructions for which you can find here `https://kubernetes.io/docs/tasks/tools/install-kubectl-macos/`

Follow the documentation at the preceding URL to get the Azure CLI set up on your system.

Once you have the Azure CLI up and running, the next step is to log in using the following command:

```
az login
```

Once you've successfully logged in, it will display the details of the subscription and tenant details for your subscription.

Run a couple of the following commands to get a feel for the Azure CLI and the basic Kubernetes commands:

- `az aks list` //: To get a list of all your aks clusters
- `az aks get-credentials --resource-group ebuy-rg --name ebuy` //: To connect to your aks cluster
- `kubectl get nodes` //: To get a list of all the nodes
- `kubectl get pods` //: To get a list of all the pods running (we don't have any pods running yet, so don't worry if you get an error message)

These are just a few commands to help you get started; if you are keen to learn about the rest of the kubectl commands head over to the official *kubectl Cheat Sheet*: `https://kubernetes.io/docs/reference/kubectl/cheatsheet/`.

Once you are happy trying out the different kubectl commands and comfortable interacting with your Kubernetes cluster, we will proceed to the next step of gathering the necessary credentials to automate deployments.

Generating credentials for your DevOps pipelines

For any DevOps pipeline to access the various resources on Azure to spin up Kubernetes clusters, it will need access permissions.

We will now collect the necessary access permissions. Ensure that you are logged in at `https://portal.azure.com`, or log in via the `az login` CLI command.

The following is a list of IDs and secrets that we need from Azure and the process to find them within the Azure portal:

- **Subscription ID**: Search for `Subscriptions` and select your subscription to display the subscription ID.

- **Tenant ID**: Search for `Azure Active Directory` and note the Tenant_ID displayed

- Then, we need to create a service principle that can create and manage resources within our resource group; we do that using the az CLI. In the terminal, fire the following command, replacing `{subscriptionid}` with the value you noted in the previous steps, and `{resource-group}` with the name of the resource group; in this case, it is `ebuy-rg`:

```
az ad sp create-for-rbac --name "MyApp" --role
Contributor --scopes /subscriptions/{subscriptionid}/
resourceGroups/ebuy-rg --sdk-auth
```

Run the command, and if all goes well, it will publish a list of configuration variables, as shown in *Figure 8.3*, which you can easily save for further use.

Figure 8.3 – Output from running the command to create a service principle

Note down the configurations from the preceding output, as we will need it in the following steps.

Now that we have all the necessary credentials we need, let's proceed to the next section on setting up the CI and CD pipelines where we will use these credentials.

Setting up CI/CD with GitHub Actions

In this section, we will learn how to go about setting up a DevOps pipeline using GitHub Actions. A DevOps pipeline is a series of steps that we define to automate the build and deployment of our apps. In this section, we will learn how to set up GitHub secrets and the workflow `.yml` file.

GitHub Actions is an automation and workflow tool provided by GitHub that allows developers to automate software development workflows and streamline their software development process. With GitHub Actions, you can create custom workflows that automate tasks such as building, testing, deploying, and releasing code directly from your GitHub repository. Other tools that we can use for CI and CD are Jenkins, Azure DevOps, Google Cloud Build, and so on. For the purpose of this chapter, we will use GitHub Actions.

Setting up GitHub secrets

As part of the CI and CD steps, GitHub Actions needs to push the Docker image to Docker Hub and spin up new Kubernetes pods, and so on. For all these activities, it needs to be able to log in to the systems with the right credentials. As a rule and for security purposes, we should never directly hardcode the usernames or passwords directly into the DevOps pipelines. The correct way is to create GitHub secrets and use those in your pipelines.

First and foremost, make sure you have committed and pushed the latest changes we've made so far to GitHub.

Let's create our GitHub secrets by first going to the **Settings** tab on the GitHub repo and then to the **Secrets and variables** section. Then, under **Actions**, we will create the following secrets along with the corresponding values that we noted down earlier from Docker and the Azure subscription:

```
AZ_CLIENT_ID
AZ_CLIENT_SECRET
AZ_SUBSCRIPTION_ID
AZ_TENANT_ID
DOCKERHUB_USERNAME
DOCKERHUB_TOKEN
```

We will create these as secrets in our DevOps pipeline. These secrets can be accessed in the pipeline as `${{ secrets.<variable-name> }}`.

Getting started with GitHub Actions

GitHub Actions is a relatively new feature provided by GitHub that allows you to create workflows to automate tasks. It can also be used to set up an automated CI and CD pipeline, which is exactly what we will use it for in this chapter.

> **Note**
> You can read more about GitHub Actions in detail here: `https://docs.github.com/en/actions`.

Creating a GitHub action is straightforward. All we need to do is, at the root of our project folder, create a folder called `.github/workflows` and then a `.yaml` file. Once pushed to GitHub, it will automatically detect that you have a workflow file and it will execute it as per the triggers:

1. Let's create our `.yaml` file at `.github/workflows/home-build-deploy.yml`, and within it, let's write the following code:

```yaml
name: home-build-deploy
on:
  workflow_dispatch:
  push:
    branches:
      - main
    paths:
      - apps/home/**
```

We will provide a name for our GitHub action; which is what will be shown in GitHub Actions. Then, we define the triggers, on `push:` and `on:workflow_dispatch`. The `workflow_dispatch` trigger allows you to manually trigger a pipeline when needed (especially when testing your pipelines), and as you can see, `on push` has further options for `branches:main` and `paths: apps/catalog/**`. This means a change to any file within the home `micro-app` that is pushed to the `main` branch will trigger this pipeline. The `paths` section is critical to ensure that the pipeline builds and deploys only the changed micro app.

2. Now, we need to define the list of jobs that GitHub actions should run; we will do this as follows:

```yaml
jobs:
  build-and-deploy:
    runs-on: ubuntu-latest
    strategy:
    permissions:
    steps:
```

For every job we define in the pipeline, we need to define what operating system the DevOps pipeline needs to run on, any strategies, what permissions to provide, and finally, the steps that it needs to run.

Now, we will expand into each of these sections.

3. Since the commands to build and deploy the micro apps remain the same, we will use a matrix strategy that allows us to define variables that can be used later in these steps. Within the strategy section, write the following code:

```
strategy:
      fail-fast: false
      matrix:
        include:
          - dockerfile: './apps/home/Dockerfile'
            image: areai51/ebuy-home
            k8sManifestPath: './apps/home/k8s/'
```

We set the `fail-fast` option to `false` so that GitHub action continues to run the pipeline for the other micro apps, even if one of them fails. Then, we define the matrix of our variables, which are as follows:

- `Dockerfile`: The path to where the micro app's Dockerfile is located in your code base

- `Image`: The path to the Docker image in Docker Hub

- `k8sManifestPath`: The location of the Kubernetes manifest files needed to spin up your micro app pod, services, and ingress

For permissions, we set the following:

```
permissions:
    contents: read
    packages: write
```

We set the `contents` scope to read and the `packages` scope to write.

The next series of steps is where the actual work happens.

As we will see, every step has two to three properties – the first is `name`; then `uses`, which is the component that is used to perform the step; and finally, `with`, which is optional and defines the additional properties required to perform the step.

All of the code in the following steps will be in the `steps:` section of the `.yml` file:

1. We start by checking out the repository:

```
- name: Checkout Repository
  uses: actions/checkout@v3.3.0
```

2. Then, we log in to Docker Hub, passing our username and the access token as the password. Note that we pass them as secrets, which we defined earlier:

```
- name: Login to Docker Hub
  uses: docker/login-action@v2
  with:
    username: ${{ secrets.DOCKERHUB_USERNAME }}
    password: ${{ secrets.DOCKERHUB_TOKEN }}
```

3. In the next step, we extract the `git SHA` value, which we will use to tag our Docker images:

```
- name: Extract git SHA
  id: meta
  uses: docker/metadata-action@v4
  with:
    images: ${{ matrix.image }}
    tags: |
      type=sha
```

4. The next step is the build and push command, where we build the Docker image by passing the micro app name via the matrix variable, and then we push that build Docker image to Docker Hub, using the `git` SHA value as the image tag:

```
- name: Build and push micro app docker image
  uses: docker/build-push-action@v4.0.0
  with:
    context: "."
    file: ${{ matrix.dockerfile }}
    push: true
    tags: ${{ steps.meta.outputs.tags }}
```

5. Once the Docker images are pushed to Docker Hub, it's time for us to set up our Kubernetes pods and services, for which we first need to set up `Kubectl`:

```
- name: Setup Kubectl
  uses: azure/setup-kubectl@v3
```

6. First, we log in to Azure using the client ID and client secrets:

```
- name: Azure Login
  uses: Azure/login@v1
  with:
    creds: '{"clientId":"${{ secrets.AZ_CLIENT_
ID }}","clientSecret":"${{ secrets.AZ_CLIENT_SECRET
}}","subscriptionId":"${{ secrets.AZ_SUBSCRIPTION_ID
}}","tenantId":"${{ secrets.AZ_TENANT_ID }}"}'
```

7. Next, we set up the Kubernetes context:

```
- name: Set K8s Context
  uses: Azure/aks-set-context@v3
  with:
    cluster-name: ebuy
    resource-group: ebuy-rg
```

8. Finally, we run the Kubernetes deploy commands:

```
- name: Deploy to K8s
  uses: Azure/k8s-deploy@v4
  with:
    namespace: "default"
    action: deploy
    manifests: |
      ${{ matrix.k8sManifestPath }}
    images: |
      ${{ steps.meta.outputs.tags }}
```

Once you've verified that all the indentation in the file is correct, go ahead and commit the file to the main branch.

Then, make a small change to any one of the code files within the home app, commit it, and push it to GitHub. After committing your change, head over to the actions tab at github. com, and you should be able to see the GitHub pipeline begin to run.

Follow the steps as GitHub Actions goes step by step through the jobs. If there are any errors, the jobs will fail, so look through the errors and make the necessary fixes. Feel free to seek help from your friends and the community as you navigate through this critical step, and keep testing until the pipeline runs successfully.

Once the pipeline builds successfully, make copies of the workflow file within the `.github/workflows` folder to build and deploy the other micro apps. We will call these files `.github/workflows/catalog-build-deploy.yml` and `.github/workflows/checkout-build-deploy.yml`.

In the respective files, change all occurrences of the word home to `catalog` and `checkout`. For example, in your `catalog-build-deploy.yml` file, you will have the following:

```
name: catalog-build-deploy
on:
  workflow_dispatch:
  push:
    branches:
      - main
    paths:
      - apps/catalog/**
```

The `matrix` section under strategies will look as follows:

```
        matrix:
          include:
            - dockerfile: "./apps/catalog/Dockerfile"
              image: areai51/ebuy-ssr-catalog
              k8sManifestPath: "./apps/catalog/k8s/"
```

Similarly, the `checkout-build-deploy.yml` file will have the following changes:

```
name: checkout-build-deploy
on:
  workflow_dispatch:
  push:
    branches:
      - main
    paths:
      - apps/checkout/**
```

Also, the `matrix` section under `strategies` will be as follows:

```
        matrix:
          include:
            - dockerfile: "./apps/checkout/Dockerfile"
```

```
image: areai51/ebuy-ssr-checkout
k8sManifestPath: "./apps/checkout/k8s/"
```

Then, make a small change, commit files to the checkout and catalog apps, and verify that only the relevant pipeline is triggered.

We can also verify the micro app pods have been successfully created within the `ebuy-ssr` Kubernetes cluster by running the following `kubectl get pods` command in the terminal.

If any of the pods don't show a ready status or have a high restart count, you can look into the pod logs using the `kubectl logs <pod-name>` command in the terminal.

With this, we have successfully created our DevOps pipeline using GitHub Actions, where we learned how to securely save our credentials as GitHub action secrets, created an individual workflow `.yml` file for each micro app, and configured it so that they are triggered only when the corresponding micro app has changed.

While these micro apps are individually running, they will not work with module federation, as the remotes on Kubernetes are different from what we ran locally. In the next section, we will update the remotes to ensure that it works on the cloud as well.

Updating the remotes

Once you have your pipelines deployed successfully, log in to `portal.azure.com`, go to the Kubernetes services, select your Kubernetes cluster, go to the Services and Ingress link, and note the external IP address for the service of the micro apps.

You can achieve the same by running the `kubectl get services` command in the terminal.

Once we have the IP address, we need to update our module federation remotes with the updated URLs.

Now, as you may have figured out, the URLs for our microapps are different locally and on Kubernetes. Since we want to be able to run our apps locally as well as on Kubernetes, we will need to conditionally load in the remotes based on whether the app is running in `dev` or `production` mode. We do this as follows:

In the `apps/home/next.config.js` file within the `remotes` object, we update the code as follows:

```
const remotes = (isServer) => {
  const location = isServer ? "ssr" : "chunks";

  const ENV = process.env.ENV;
```

```
  const CATALOG_URL_LOCAL = 'http://localhost:3001';
 const CHECKOUT_URL_LOCAL = 'http://localhost:3002';

  const CATALOG_URL_PROD = 'http://<your-k8s-ip-address>'
  const CHECKOUT_URL_PROD = 'http://<your-k8s-ip-address>'

  const CATALOG_REMOTE_HOST = ENV === 'PROD' ? CATALOG_URL_PROD
: CATALOG_URL_LOCAL;
  const CHECKOUT_REMOTE_HOST = ENV === 'PROD' ? CHECKOUT_URL_
PROD : CHECKOUT_URL_LOCAL;

  return {
    catalog: `catalog@${CATALOG_REMOTE_HOST}/_next/
static/${location}/remoteEntry.js`,
    checkout: `checkout@${CHECKOUT_REMOTE_HOST}/_next/
static/${location}/remoteEntry.js`,
  };
};
```

What we do here is define a new variable called ENV that captures whether the app is running in dev or prod mode, then we create the consts for LOCAL URL and PROD URLS for our micro apps, and conditionally set the values of the CATALOG_REMOTE_HOST and CHECKOUT_REMOTE_HOST values based on the ENV values.

Make the same set of changes to the next.config.js files in the checkout and catalog apps, and then save the changes.

Now, we can and build the apps locally to verify that things work fine.

Run the pnpm dev command from the root of the project.

Once this works locally, let us commit the changes to Git and let the GitHub actions auto-trigger and deploy the new apps to our Kubernetes cluster.

Once it's all done, head over to the URL of the home micro app (http://<your-k8s-ip-address>/) and verify that the app is working.

> **Important note**
>
> Make sure the catalog and checkout apps are deployed first before the home app pipeline starts. This is because, in prod mode, the home app now expects the remoteEntry.js files to be present at the URLs we defined in the CATALOG_URL_PROD and CHECKOUT_URL_PROD constants.

Summary

And with that, we have come to the end of this chapter. I hope you have been able to follow along and enjoyed the joys and pains of wearing a DevOps engineer's hat.

As you can see, we covered a lot in this chapter. We learned about Kubernetes and its various key components. We saw how you spin up an empty Kubernetes cluster on Azure and learned about the Kubernetes spec files that deploy our micro apps into a Kubernetes cluster. We learned how to containerize our micro apps using Docker and how to set up Docker Hub as a remote image repository. Then, we went through the detailed steps of setting up a CI/CD pipeline using GitHub Actions, and finally, we made the necessary tweaks to our code base so that we can run our module-federated microfrontend on Kubernetes. Now that you have managed to complete this chapter, give yourself a pat on the back and take a well-deserved break before we start with the next chapter, where we will see how to manage our microfrontend in production.

Part 4:
Managing Microfrontends

This part focuses on managing microfrontends in production, including versioning, feature toggles, rollbacks, and branching strategies. It provides guidance on operations best practices.

This part has the following chapters:

- *Chapter 9, Managing Microfrontends in Production*
- *Chapter 10, Common Pitfalls to Avoid When Building Microfrontends*

Managing Microfrontends in Production

Being able to develop and test web applications on your local computer is great; however, deploying them them to production, maintaining them, and releasing new features while your applications are being visited by hundreds and thousands of visitors takes your software development skills to the next level. This chapter will cover some of the key concepts around deploying and maintaining your microfrontends in production.

In this chapter, we will cover the following topics:

- Branching strategies
- Versioning
- Rollback strategies
- Feature toggles

By the end of this chapter, you will have taken the first steps toward reliably maintaining your microfrontend applications in production.

Foundational components for a strong software delivery model

When it comes to deploying and maintaining applications in production, I recommend using **DevOps Research and Assessment's** (**DORA's**) software delivery maturity model to help prioritize areas to focus on and aspects of your production deployment processes you should optimize.

> **Important note**
>
> The software delivery maturity model talks about four key areas—namely, *Deployment frequency*, *Lead time for changes*, *Time to restore service*, and *Change failure rate*, and these are categorized as *Elite*, *High*, *Medium*, and *Low*. You can read more about this in detail here in the *State of DevOps 2021* report: `https://dora.dev/publications/pdf/state-of-devops-2021.pdf`. You can also register and view the rest of the reports at `https://dora.dev/publications/` and `https://cloud.google.com/devops/state-of-devops`.

We will look at a couple of key components that help you create the right foundation to ensure you are able to move up the maturity model as your team gains more confidence in deploying and managing microfrontends in production.

Branching strategies

In my opinion, the branching strategy is the most critical component that helps you improve the *Deployment frequency* and *Lead time for changes* metrics.

GitFlow and GitHub Flow are two popular branching strategies for Git-based version control systems, each with its strengths and weaknesses.

GitFlow is a branching model that uses two long-lived branches, `main` and `develop`, as well as feature, release, and hotfix branches.

GitHub Flow, on the other hand, is a simpler and more flexible branching strategy that is suitable for smaller teams and projects. It revolves around a single main branch, typically `master` or `main`, and encourages developers to make changes in feature branches that are then merged into the `main` branch through pull requests.

In our opinion, when working with microfrontends and monorepos, GitHub Flow is the only viable branching strategy. This is primarily because of the following reasons:

- **Simplicity**: GitHub Flow simplifies the development process by enforcing a single, linear history on the `main` branch. Every feature, bug fix, or improvement is developed on a separate branch created from `main`, and once ready, it's merged back into `main`. There are no long-lived branches apart from `main`, avoiding the complications of managing, syncing, and maintaining multiple long-term branches.

- **Isolation of changes**: In a monorepo, it's crucial to ensure changes to one project don't unintentionally impact another. GitHub Flow's practice of isolated branches for each new feature or bug fix aids in containing the scope of changes, reducing the risk of cross-project interference.

- **Continuous Integration/Continuous Deployment (CI/CD)**: GitHub Flow is designed with CD in mind. In monorepos, this can be even more beneficial. Since all projects live within the same repository, it's easier to ensure that all changes are tested and deployed consistently.

- **Reduced merge conflicts**: With a strategy such as Git Flow, where changes are often merged into `develop` or `release` branches before `main`, there can be significant delays between when code is written and when it's deployed. In a fast-paced monorepo, this can lead to complex merge conflicts. GitHub Flow mitigates this by encouraging frequent merges directly into `main`.

While working with GitFlow we've also seen that it is beneficial to slightly switch how code is merged to `main` based on whether you are in active development or post your first production deployment.

During active development

During the active development phase of a project, team members diligently generate feature branches, merging them back into the `main` branch only after the pull requests have received the necessary approvals. It's common practice, and possibly part of your CI builds, to initiate an array of automated unit tests on each pull request prior to merging into the `main` branch. Additionally, a suite of integration and end-to-end tests are preferably executed on the `main` branch each night. This routine helps ensure that any disruptions in the `main` branch are identified and rectified promptly.

Let us look at the branching and merging workflow with GitHub Flow during active development:

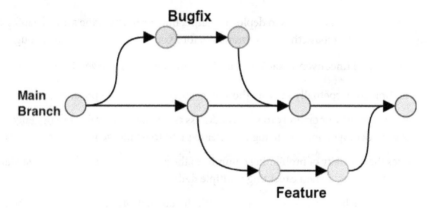

Figure 9.1 – Branching and merging strategy during active development

As you can see, with GitHub Flow, branching and merging is quite straightforward. Developers branch off the `main` branch and merge back into `main`.

After the first release to production

Subsequent to your inaugural production release, the merging strategy undergoes a subtle evolution. The teams continue to create feature branches off the `main` branch as before; however, the process diverges post-feature testing and approval. The tested feature release is rebased with `main`, tagged and deployed directly from the feature branch into production. It is only once stability in production is confirmed that the feature branch is merged into the `main` branch.

The following workflow will help illustrate the process:

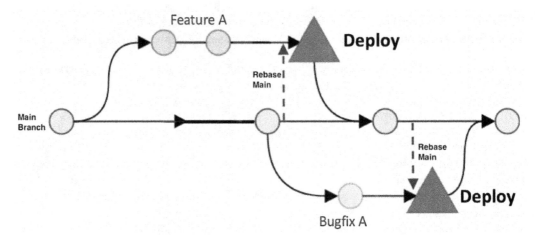

Figure 9.2 – Branching and releasing after the first release

As you can see in *Figure 9.2*, the process to deploy a feature, bug, or hotfix is the same, keeping things simple. A key step here is to rebase the main branch before you deploy your feature or bug.

Although this may sound unconventional, adhering to this approach has several benefits, as follows:

- The main branch perpetually mirrors the stable, current version in production.

- The team is given the opportunity to swiftly address and resolve any minor issues encountered during feature deployment, stabilizing the release prior to its merge into the main branch.

- It eliminates the necessity of prohibiting commits to the main branch until the release stabilizes, an impractical strategy when executing multiple daily deployments.

- As the main branch is always in alignment with the current production version, the process of deploying a feature or hotfix remains consistent.

- Since all merges into the main branch are post-release, the likelihood of a disrupted main branch is significantly reduced. Such disruptions can impede a large development team and halt further production deployments until the issue is resolved.

In the context of GitHub Flow, it is vital to note the following:

- Prior to deploying your feature branch, ensure a final rebase from the main branch. This ensures that your feature branch encompasses all previous deployments executed while your feature was in development.

Contrary to popular belief, GitHub Flow development is not exclusive to small 2-3-member teams. It is, in fact, notably advantageous for large teams operating in small, focused squads or pods.

Versioning micro apps

The versioning of applications being deployed to production is standard practice, and there are different ways of defining versioning strategies. Versioning is important from many aspects; it helps in managing changelogs and mapping the different features and bugs to a build. It also helps with rollback strategies.

Semantic versioning, also known as **SemVer**, is a popular technique to define versions. It follows the format of `MAJOR.MINOR.PATCH`.

This structure not only helps users to understand the nature of changes in the new release but also helps with dependency management in software systems.

Our recommended strategy is to use SemVer as a guide. Each micro app should adhere to its own versioning rules, ensuring that any changes to the application's public features are reflected in its version number. A micro app can increment its version number following the `MAJOR.MINOR.PATCH` pattern each time there's a major feature release, minor feature release, or a bug fix change. We recommend prefixing the app name to the semantic version—for example, `catalog-2.8.3` would mean the following:

- **Micro app name**: `catalog`
- **Major version**: 2
- **Minor version**: 8
- **Path/bug fix version**: 3

For a simple way to manage tagging, whenever you are working on a release, we recommend creating a release branch such as `releases/catalog-2.0.1`.

Once the release is tested and ready to deploy, we tag it like so:

- `git tag catalog-2.0.0`
- `git push origin catalog-2.0.0`

With a CI/CD pipeline, we can set it to automatically trigger a deployment to the test environments whenever a new tag is detected.

Note that while SemVer's official definition uses the term *breaking changes* to define a major version, with microfrontends and module federation we can't really have a breaking change. Hence for us, a micro app containing a major feature release would warrant a major version number increment.

A common scenario with multiple micro apps and multiple frequent releases is that it becomes challenging for everybody in the team to know which version of which micro app is currently in production. A simple way to solve this is to have a `/versions` route in every micro app that will display information such as the current version number, the date of release, the branch it was released from, and so on. This can be super helpful for developers trying to debug issues in production.

Here is an example of information we had on a /versions route:

```
{
appName: Catalog
branchName: release/catalog-2.0.0
tagName: catalog-2.0.4
deployedDate: Thu 25 May 2023 13:34:04 GMT
}
```

For a large number of micro apps or frequent releases, manually tagging every version might become tedious. You might want to consider automating this using a script. Here is an example of a bash script:

```
# Microfrontend names
MICROFRONTENDS=(home catalog checkout)
for i in ${MICROFRONTENDS[@]}
do
    cd apps/$i
    # Fetch latest tags
    git fetch --tags
    # Get latest version from Git
    VERSION=$(git describe --tags `git rev-list --tags --max-
count=1`)
    # Increment version
    npm version patch
    # Add release notes
    git commit -am "Release v$VERSION [skip ci]"
    # Tag commit
    git tag v$VERSION
    # Push changes
    git push --follow-tags origin main
    # Build microfrontend
    pnpm run build
  done
```

In the preceding code, you will notice that the script loops through each micro app in the apps folder, fetches the latest tags from git using the git describe and git rev-list commands, runs the npm version command to update the version numbers, and then commits and pushes the updated tags back to git.

We can also use tools such as `semantic-release` or `standard-version`. These tools automate version management and changelog generation based on commit messages.

Versioning and tagging micro apps is critical to ensure that there is clarity with all stakeholders on the current status of which version of each micro app is currently in production. As we will see in the next section, it also plays a critical role in rollback strategies.

Rolling back a micro app

A rollback strategy is a key component to managing any production software. This impacts the *Time to restore service* metric.

Rollback strategies for microfrontends center on the ability to revert a specific micro app or the entire system to a previous stable state when issues arise during or post-deployment. *Thanks to the independence of microfrontends, a rollback doesn't necessarily affect the entire application but can be targeted to the problematic component, reducing overall system disruption.*

The simplest rollback strategy involves utilizing version control systems such as Git along with CI/CD pipelines. In this setup, each microfrontend has specific tagged releases, which are stored and can be redeployed if required. For instance, if the current version of a microfrontend is `catalog-1.2.3` and an issue is detected, you can quickly revert to the previous stable version, `catalog-1.2.2`, by triggering the corresponding deployment in your CI/CD pipeline.

Additionally, leveraging a blue-green deployment strategy can be effective. In this approach, two environments—blue and green—are maintained. While one serves live traffic (blue), the other (green) is idle or being prepared for the next release. If something goes wrong with the green environment post-deployment, you can quickly switch back to the blue environment, effectively rolling back the changes.

Rollbacks in Kubernetes are straightforward thanks to its declarative nature and built-in versioning mechanism. When a new deployment is created, Kubernetes automatically versions it and stores its details. If an issue arises with a new release, you can quickly roll back to a previous version using the `kubectl rollout undo` command. For instance, if you find a problem with a deployment named `deployment/catalog`, you can roll back using the `kubectl rollout undo deployment/catalog` command. Kubernetes will revert the deployment to the previous stable version gracefully without any downtime, making it a powerful tool for managing rollbacks in microfrontend architectures.

When rolling back a micro app, it is important to be aware of any incompatibilities with backend APIs and whether the corresponding backend API also needs to be rolled back.

Rollbacks at times can be painful, and the need for rollbacks can be mitigated by releasing new features or versions of micro apps using feature toggles, which we will see in the next section.

Deploying micro apps with feature toggles

Feature toggling, also known as feature flagging, is a powerful technique that allows individual features to be turned on or off at runtime without requiring a redeployment. This is particularly useful in a microfrontend architecture, as it enables the independent release and control of micro apps across multiple micro applications.

With feature toggling, teams can deploy new features to production but have them "hidden" behind a toggle until they're ready to be released. This allows for extensive testing in the live environment and enables progressive delivery techniques such as canary releases or A/B testing. If any issues arise with the new feature, it can be quickly "switched off" via the feature toggle, effectively mitigating the impact without requiring a full rollback or redeployment.

Unleash (`https://www.getunleash.io/`) is a popular open source tool for feature toggles.

Feature toggles can be used to provide different experiences for different users. For instance, you can use them to selectively enable features for specific user groups, such as beta testers or premium users.

However, feature toggling needs to be managed carefully to avoid an accumulation of outdated toggles, which can lead to code complexity and technical debt. Regular audits and cleanup of feature toggles should be part of the development process.

With this, we come to the end of this section, which covered some of the foundational elements of managing your microfrontends in production. This goes in conjunction with everything we saw in the previous chapters about deploying microfrontends to the cloud and eventually helping reduce the overall stress involved with deploying and maintaining applications in production.

Summary

As we conclude this chapter, let us quickly summarize what we've learned so far. We learned about DORA's software delivery performance metrics: *Deployment frequency*, *Lead time for changes*, *Time to restore service*, and *Change failure rate*. We then had a look at some of the foundational elements that teams need to focus on to ensure they are set up for success.

We learned about branching strategies and that GitHub Flow is the preferred branching strategy. We also learned about the nuances of workflows when software is being built versus when it's deployed.

We learned about the right way to version our micro apps. We also learned about the importance of rollback strategies and how microfrontends help minimize the blast radius. And finally, we learned about feature toggles and how we can gradually release new micro apps into production via feature toggles and, more importantly, if there are any problems.

In the next chapter, we will look at some common pitfalls to avoid when building microfrontends.

10

Common Pitfalls to avoid when Building Microfrontends

We've come a long way! We've learned how to build microfrontends, how to deploy them to the native cloud, and how to manage them in production.

As we start working with microfrontends, we will make mistakes, but we will learn from them and eventually build our own set of best practices, discovering what works best for our use cases. However, it is always a smart thing to learn from others' mistakes as well. In this chapter, we will cover some of the pitfalls earlier teams faced when working with microfrontends.

We will teach you about some common pitfalls and how to avoid them, which are as follows:

- Not making your microapps too small
- Avoiding the overuse of common shared code/libraries
- Avoiding multiple frameworks within a microfrontend
- The inability to deploy individual micro apps
- Excessively relying on state
- Avoiding build-time compilation to assemble Microfrontends
- Avoiding packing your micro apps into NPM packages

By the end of this chapter, you will have learned about the various pitfalls developers fall into when transitioning from single-page apps to microfrontends.

Don't make your micro apps too small

We touched upon this at the start of the book, but it's important to stress it again. Way too many developers think that, in a microfrontend architecture, the micro apps need to be really small. This is not true, as creating very small microapps greatly increases the complexity and maintenance headaches, without achieving any benefits.

In trying to identify what the right size is for your micro app, we've seen it helps if we take into consideration the following points:

1. Is it the largest possible micro app that can independently exist?

2. Is it the largest possible micro app that's owned by a single agile scrum team?

3. Does this app undergo changes and updates that are at a pace different from the rest of the application?

4. Another point to consider is thinking in terms of domains, based on domain-driven design principles, to determine what business features a given micro app should support or not support.

If your answer to all the preceding questions is yes, then the micro app is the right size. If the answer is no to any one of the preceding questions, then either we haven't broken down our micro apps in the right way or microfrontends may not be the right architectural choice.

Another guide to help identify the right size for your app is to look at the atomic design pattern (`https://bradfrost.com/blog/post/atomic-web-design/`), which defines how components are structured in an application.

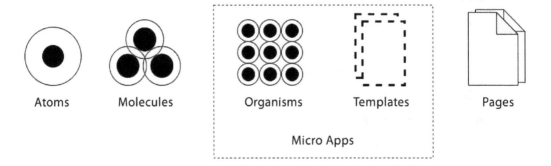

Figure 10.1 – Organisms and templates can be converted to micro apps

If you look at the atomic design pattern in *Figure 10.1*, the ideal level to break down your app into micro apps would be either at the organism level or the template level; anything other than that would be either too small or too big.

Breaking down the application into the right-sized micro app is key to building a performant and scalable microfrontend architecture, and investing more time in getting this right will pay high dividends as we move forward.

Avoiding the overuse of Shared Component Code

When it comes to building microservices or microfrontends, team independence is the highest priority. Anything that makes a team dependent on another team should be strongly discouraged.

In our experience as software developers, we've always come across principles such as **reusability**, **Do not Repeat Yourself** (**DRY**), and so on. In fact most senior developers are constantly looking how do they create common utilities, helpers shared components, and so on, to help the teams be more productive.

However, when it comes to the world of microservices and microfrontends, overuse of these shared libraries can lead to what is called "dependency hell" or a "distributed monolith," which is the worst of both worlds.

This is bad for microfrontends because using shared libraries or code immediately takes away the independence of teams, as now two or more teams are dependent on updates or bug fixes to be made for this shared library, in order for them to be able to proceed further.

As more and more teams start using a shared library, it tends to start getting bulkier, as it now needs to accommodate the use cases of the different teams. There is also a constant risk that changes or updates to this shared code may break the functionality of one or more teams.

Hence, when it comes to microfrontends, we need to be strict about not falling into this trap. As a rule of thumb, we should avoid creating any business or application logic as shared common code. One item that can ideally be shared between micro-apps is the UI component library because we want to ensure that all micro apps have a consistent look and feel. Another item that can be put into a shared library is any other low-level utility function that doesn't contain any business logic. Some examples of these would be an HTTP client, an error-handling utility, or other utilities to format dates or manipulate strings.

Remember that with monorepos, it's a lot easier to "find and replace" than to deal with the challenges of a distributed monolith.

While initially the whole idea of prioritizing team independence over code reuse may sound like an anti-pattern and not a smart thing to do, speaking from experience, this is the second most important point to keep in mind when you want your teams to move fast and frequently deploy code to production.

Avoiding using multiple frameworks in your microfrontend

One of the benefits of microfrontends is that, technically, it's possible to have each app built using a different framework. However, just because it's possible doesn't mean you have to. There are numerous drawbacks to using multiple frameworks within a single microfrontend:

- The cognitive overload for team members as they potentially switch from one team to the other over time is very high.

- Since every framework comes with its own JavaScript bundle, and since every framework will have a different set of NPM modules that the team uses, the amount of JavaScript code transferred to the user's devices will be high. Therefore, we will not be able to take full advantage of browser caching or service worker caching, since each app uses its own bundle.

- Different frameworks will have different performance challenges and issues, and each team will have to deal with them individually and not be able to use the collective knowledge within the broader team.

Having said that, it is fine to have multiple frameworks or multiple versions of them for a short transitional phase when you evaluate a new framework or incrementally upgrade to a newer version. Overall, though, having multiple frameworks as an architecture principle should be avoided.

An inability to deploy an individual micro app

One of the primary reasons to adopt a microfrontend architecture is to allow certain parts of an application to be independently updated without impacting the rest of it.

This obviously means that we need the ability to build and deploy each micro app independently. If your DevOps build and release pipeline can't do this, then it's better to go with **Single-Page Application (SPA)** architecture.

In the past, many DevOps tools weren't sophisticated enough to work with monorepos or microfrontends; however, most of the latest tools are better equipped to detect which folders have changed and only trigger the necessary app builds.

Hence, when working on a microfrontend architecture, it is critical that you've thought it through and through, including how it will be deployed, as this will impact the choice of tools you select for the DevOps pipeline or the monorepo.

For example, if your DevOps pipelines can be conditionally triggered based on which micro app has changed, then you are free to choose any monorepo tool.

However, if your DevOps pipeline is unable to detect changes, or if you are limited to a single pipeline for all your microfrontends, then going with a mono repo tool such as Nx, which has built-in change detection, would be more suitable.

Excessively relying on state

With the advent of React, state management became a thing, and with it rose the popularity of tools such as Redux that advocated a single central data store to manage state. Over time, developers seem to have become obsessed with state management, relying far too much on these state management libraries. When developers make the shift from SPAs to microfrontends, they continue their obsession with state and spend a lot of time trying to persist state, making it work across different micro apps. With SPAs and also microfrontends, it is important to sparingly use these application-level states. When working with microfrontends, we encourage exploring concepts around Pub/Sub or an event emitter approach to sharing data between different micro apps. Alternatively, look at native browser data stores, such as session storage, IndexedDB, or local storage to manage persistent state, or if none of these is an option, then explore lightweight state management libraries such as Zustand or React's Context API.

As you may have realized by now, when building microfrontends, there is a fair bit of unlearning and relearning involved, especially if you have been building SPAs for a while. The use of state management in microfrontends is something that needs to be understood and is also the most difficult change that some developers occasionally have to deal with, especially those who have got used to excessively relying on state.

Avoiding build-time compilation to assemble Microfrontends

There is a current trend in the frontend community to move as many tasks as possible to the build time phase of application compilation, rather than the runtime. Good examples of these are static site generation, where the HTML pages are generated at build time, or **Ahead of Time** (**AoT**) compilation in Angular, which improves the overall performance of an application.

While, in general, build-time compilation is a good practice, reducing the load on the browser and JavaScript engines during the runtime phase, it doesn't help when assembling the microfrontend. This is because every time any microfrontend changes, you need to rebuild the assembly layer as well, defeating the principle of independent micro app deployments.

We can choose to have individual micro apps do more work during build time (e.g., generate static pages), but the assembling of micro apps or module federation should always be done on the server or at runtime.

This is another key point to keep in mind to ensure we don't blindly follow "popular trends." It is important to always remember what the key principles of your architecture pattern are and that you've thought through your pattern, end to end and all the way to how it will be deployed into production.

Avoiding packing your micro apps into NPM packages

Another common trend within the SPA world is to convert any sharable modules into NPM packages for easier distribution and then import them into other apps.

In our experience, we have seen a few teams package and version their micro apps into NPM modules before importing them into the host or assembly app. We strongly discourage this practice for the primary reason that every time a new version of a micro app is published as an npm module, all the hosts using that micro app will need to update their `package.json` files and rebuild and redeploy their apps, defeating the primary principle of independent deployments. We covered this in a bit of detail in *Chapter 2, Key Principles and Components of Microfrontends*, in the *Prefer runtime integrations* section.

Summary

With this, we come to the end of this chapter. Being a relatively new architecture pattern, the concepts and best practices around microfrontends are constantly evolving.

In this chapter, we saw some of the common pitfalls that teams have fallen into while building microfrontends – namely, things such as not being able to identify the right level at which to break down an app into a micro app, overuse of state management libraries, using multiple frameworks within a micro app, the inability to individually deploy a micro app, overuse of shared common code, and ending up with a build-time integration. Hopefully, this chapter will prevent you from repeating the same mistakes your peers have made in the past.

Another important point to remember is to understand the reasoning behind these best practices, looking at them through the lens of your specific use case. Follow the best practices that apply to your use case and tweak the ones that don't quite fit it.

As the famous saying goes, "*The answer to every architecture question is… it depends.*"

In the next chapter, we will look at some of the emerging trends in the world of microfrontends that you should keep an eye on.

Part 5: Emerging Trends

This part explores the microfrontend landscape and analyzes bleeding-edge techniques and technologies applicable to microfrontends, such as generative AI, edge functions, and the island architecture pattern.

This part has the following chapter:

- *Chapter 11, Latest Trends in Microfrontends*

11
Latest Trends in Microfrontends

The world of frontend engineering is constantly evolving, and as we go about building microfrontends following the currently available tools, approaches, and best practices, it is important to keep an eye on the latest trends that are evolving in this space and keep exploring and experimenting with them to see how they can help us become more efficient and build better apps.

In this chapter, we will cover some trends that can influence how we build microfrontends in the future. Some of the trends we will explore are the following:

- The name *microfrontends* itself and what is a better term for it
- The island pattern of mixing static content with dynamic content
- Looking at other build tools beyond Webpack
- WebAssembly
- Cloud or edge functions
- How generative AI can influence our work

By the end of this chapter, we will have learned about the latest trends in the frontend engineering space that impact how we build microfrontends.

Microfrontends – decoupled modular frontends

The term *microfrontends* has obviously become very popular, and this entire book uses it, but to be honest, I've always felt it was poorly coined and unfortunately, it has stuck within the community. As mentioned a couple of times, the word *microfrontend* has led to a lot of misinterpretation, leading to bad architectural patterns that cause more harm than good. A new proposal has been put forward to start calling them **composable decoupled frontends** (https://microfrontend.dev/), which I think is apt and clearly explains the intent and purpose of what we are building. I really hope the community starts picking this term up and that we collectively all start moving to building and calling microfrontends what they really are and defining what they are really supposed to do.

I'm sure many of you will wonder how simply changing the name helps and what's really in a name; however, I feel that, in this case, a name that clearly articulates the architecture pattern greatly reduces the misconceptions, misinterpretation, and complications arising from wrongly architected systems. As you will have realized through the course of this book, it is all about building modular applications that are decoupled from each other and hence they should be rightfully called **Decoupled Modular Frontends**.

The island pattern

Statically generated pages are gaining a lot of popularity as they ship very little to no JavaScript; however, the challenge with them has always been on how to serve dynamic content.

The **island pattern** aims to solve this problem. It was made popular by the Astro build framework, wherein we have our application published as a set of statically generated HTML pages, within which the dynamic parts of the page are imported as islands.

Here is an example of how this can be achieved using Astro, a popular framework for building statically generated sites.

You can read more about this at `https://docs.astro.build/en/concepts/islands/`:

```
//index file
---
// Example: Use a dynamic React component on the page.
import MyReactComponent from '../components/MyReactComponent.jsx';
---
<!-- This component is now interactive on the page!
     The rest of your website remains static and zero JS. -->
<MyReactComponent client:load />
```

Run the Astro build command, test the app locally, and look into your Inspect command; you will notice that while the rest of the page is plain HTML with little to no JavaScript, MyReactComponent is a small JavaScript element and executes on the client side.

As you can see, with the island pattern, we get a clear distinction between static and dynamic content with the potential added benefit of not being locked down to a single framework for all parts of the application.

Having said that, there are a few differences between the island pattern and microfrontends, including the following:

- Islands in Astro are components that are hydrated/rendered on the client side, while microfrontends are independent applications with their own code bases, routing, and backends. Microfrontends are more isolated and decoupled.

- Astro builds the entire app and islands at build time. Microfrontends are built and deployed independently. Astro has a unified build, while microfrontends can have separate builds.

- Routing in Astro happens in the shell, while each microfrontend manages its own routing. Astro islands don't have independent routing.

- Astro islands can communicate with each other via Astro integration, while microfrontends typically communicate via well-defined APIs and events. Islands have tighter coupling and integration with the Astro app.

Beyond Webpack with ES Modules

With the dawn of JavaScript-based frameworks, Webpack rose in popularity, and it became the de facto module bundler for all JavaScript frameworks. However, bundling/compiling large applications with Webpack can be very slow, and manually configuring it to efficiently bundle an app is very complex. Recently, a new breed of bundler tools that takes advantage of ES modules has taken the frontend world by storm, promising compilation over 20 times faster than Webpack.

ES modules are a standardized way to define and import modules in JavaScript. They allow for modular code organization, which can make it easier to develop and maintain large applications. ES modules also provide a clear and explicit syntax for importing and exporting code, making it easier to reason about the dependencies between different modules.

Each of our micro apps can be exported as ES modules, and by using dynamic imports, we can embed them into our host application.

The entire microfrontend application can be bundled using an ES build-based module bundler such as Vite (`https://vitejs.dev/`). Monorepo frameworks such as Nx allow you to easily configure using Vite as your module bundler.

We can scaffold out a React app using Vite as follows:

```
pnpm create vite microfrontend-app --template react
```

Here is a rough example of how this can be achieved:

```
// Catalog App
function CatalogApp() {
  return <h1>Hello World</h1>;
```

```
}
export default CatalogApp;
```

In the host app, we use the classic React `suspense` and `lazy` functions to load in `CatalogApp` at runtime:

```
// Host App
import React, { lazy, Suspense } from 'react';

const CatalogApp = lazy(() => import('./catalog'));

function App() {
  return (
    <>
      <Suspense fallback={<div>Loading...</div>}>
        <CatalogApp />
      </Suspense>
    </>
  );
}
```

As you will have noticed, we have managed to get our app working without using Webpack or Webpack's module federation, and I'm sure you will also notice how fast the app builds after any changes that you make.

We believe ES modules and ES build systems will soon replace Webpack to become the de facto tools of choice for building all modern frontends. What is also interesting to note is that while React's `lazy` and `suspense` functions are commonly thought of as performance optimization techniques, we take advantage of their ability to load modules in real time to build microfrontends.

Using WebAssembly Modules

WebAssembly (**Wasm**) has been around for many years now. Despite its huge benefits in terms of performance and low bundle size, it hasn't gained much popularity, primarily because it wasn't easy for developers to build a WASM module. However, now that people are starting to work with tools such as Rust, it gets fairly easy to build WebAssembly modules with Rust. We anticipate that WebAssembly will become mainstream when building applications that require a high level of computation on the browser.

WASM modules can work really well in a microfrontend architecture, where the critical compute-intensive modules are built in WASM wrapped as a micro app and imported into a microfrontend architecture in which the rest of the micro apps in the microfrontend are built using the standard React.

Here is a rough approach of how you could set this up in your module federated Next.js app. Use our module federation code from *Chapter 6*. First build a Rust app using wasm_bindgen within a /rust folder.

To compile the rust app to wasm we need to install the wasm-pack-plugin as using pnpm install @wasm-tool/wasm-pack-plugin and use it in the next.config.js configuration as follows:

```
const NextFederationPlugin = require("@module-federation/
nextjs-mf");
const WasmPackPlugin = require('@wasm-tool/wasm-pack-plugin');

const path= require("path")

const remotes = (isServer) => {
  const location = isServer ? "ssr" : "chunks";
  return {
    catalog: `catalog@http://localhost:3001/_next/
static/${location}/remoteEntry.js`,
  };
};
module.exports = {
  webpack(config, options) {
    config.plugins.push(
      new WasmPackPlugin({
        crateDirectory: ('./rust'),
    }),
      new NextFederationPlugin({
        name: "catalog",
        filename: "static/chunks/remoteEntry.js",
        exposes: {
          "./Module": "./pages/index.tsx",
        },
        remotes: remotes(options.isServer),
        shared: {},
        extraOptions: {
          automaticAsyncBoundary: true,
        },
```

```
    })
  );
  config.experiments = {
    syncWebAssembly: true,
  };
  config.module.rules.push({
    test: /\.wasm$/,
    type: 'webassembly/sync',
  });

  return config;
  },
};
```

Then using dynamic imports, import the wasm module into the index page of the remote app. And finally using the approaches we used in *Chapter 6* import the remote app into the host app.

WASM is already being used in some very popular web-based tools such as Figma, AutoCAD, Google Earth, the Unity game engine, and so on. Combining WebAssembly modules with microfrontends helps bring the best of both worlds: the power and performance of WASM, and the ease of use and modularity of microfrontends.

Edge Functions or Cloud functions

Edge functions are gaining a lot of popularity, as they provide the power to compute on the edge. Think of them like a **Content Delivery Network** (**CDN**) but with the power and ability to run computations.

The primary benefits of edge functions are that they provide very low latency, which greatly helps improve performance, and they use an automatic distributed deployment, which mitigates single points of failure and helps improve scalability.

Edge functions and microfrontends work quite well hand in hand, where you can have each micro app deployed within a cloud function; this automatically allows for modular deployments, and each team can manage its cloud functions independently.

Cloudflare is one of the most popular providers that support cloud functions. Cloudflare Workers and most recently Cloudflare Pages support computing on the edge. Here is an example of how to deploy a Next.js App on Cloudflare Pages using Edge Runtime.

1. Start with any of the existing Next.js apps we've built.

```
npm install --save-dev @cloudflare/next-on-pages
```

2. Commit your changes and push them into a Git repo.

3. Login into the Cloudflare dashboard and go to **Workers & Pages | Create Application | Pages | Connect to Git**.

4. 4. Select the repo where you pushed the code and in the Setup builds and deployments, select Next.js as your Framework. Leave the rest of the settings as default.

Next we need to set the Compatibility Flags which we do by going into the **Pages | Settings | Functions | Compatibility Flags**. And we need to set the value to `nodejs_compat`.

From the Deployment Details section go to the **Manage Deployment** and select **Retry deployment** from the dropdown.

Thanks to the low costs and ease of deployments, we believe there is a great potential to deploy all frontend applications, irrespective of whether they are microfrontends or not, on platforms such as Vercel, Cloudflare, Fastly, and so on.

Most edge function providers have very good support for the JavaScript ecosystem; however, it is important to keep in mind that based on the vendor/platform you are working on, there may be certain restrictions. For example, Cloudflare limits the size of each worker to be under 1 MB, or it explicitly supports package versions that are compatible with the broader Node.js runtime environments. For Cloudflare, you can read more about Node.js compatibility here: `https://developers.cloudflare.com/pages/framework-guides/`.

Generative AI and Microfrontends

Generative AI has clearly taken the world by storm. We are seeing amazing examples of generative AI being able to generate complete end-to-end applications.

When it comes to building microfrontends, it will be very interesting to see how things evolve. While I believe generative AI can't take over a developer's job, I do see interesting use cases of how generative AI can work hand in hand with microfrontends in building unique customer experiences.

Generative AI can be leveraged to dynamically generate and assemble various parts of a web application. By intelligently analyzing user behavior, preferences, and real-time context, AI can create microfrontends that are tailor-made for individual users, resulting in a highly personalized and optimized user experience. This approach also simplifies the development process by allowing developers to focus on creating modular, composable micro apps, while the AI system takes care of the overall assembly and rendering of the web application.

New AI-powered tools such as GPT-Engineer, smol-ai, and Auto-GPT are emerging, which allow developers to describe application requirements using plain text or Markdown. These tools then scaffold and generate code for the full application based on the developer's specifications. This removes the need for manually writing all of the code, and instead, lets the AI handle much of the initial setup. These kinds of AI developer assistants are still at quite an early stage; developers will need to learn skills such as crafting effective prompts to get the most consistent and accurate results from the AI, but the potential is there for AI to significantly enhance and accelerate development workflows. The key will be continuing to improve the AI's code generation abilities while also helping developers provide the right input and guidance.

The use of AI in microfrontends can lead to more efficient resource utilization and improved performance, as the system can adaptively load and unload components based on user interactions and needs. This innovative integration of AI and microfrontends has the potential to revolutionize the way web applications are designed, developed, and delivered to users.

Summary

With this, we have come to the end of this chapter and the book. We really hope you've enjoyed the journey.

In this chapter, we looked at a few new trends that will influence the way we build and deploy microfrontends. We saw how concepts such as the island pattern can help interlace dynamic content blocks within a statically generated multipage app. We saw how the new Rust-based bundler can be many times faster than Webpack. We learned about WebAssembly and how it can be used within microfrontends, and finally, we looked at cloud functions, which have the potential to become the default solution for deploying all modern frontend applications.

I'm truly excited about how quickly technology is evolving and how it affects the way we build our applications. I can't wait to see you go out in the wild and build things that make this world a better place.

In closing, it is essential to remember that the world of microfrontends, much like our dynamic digital landscape, is in a constant state of evolution. The concepts, techniques, and technologies we have unraveled throughout this journey, such as Module Federation and the intriguing practice of deploying microfrontends to the cloud, are just the beginning of this ever-evolving tapestry. They provide us with the building blocks to construct high-performing, scalable, and maintainable frontend architectures. Yet, the future beckons with promises of newer trends and advancements that will continue to redefine the horizon.

I encourage you, the next generation of developers, to step into this exciting journey and build upon the foundational knowledge this book has attempted to provide. Challenge the status quo, experiment with the latest trends, and mold them to fit the unique demands of your projects. It's a grand time to be a frontend engineer, and the world awaits the innovative solutions you will create using React and microfrontends. Remember, every line of code you write is an opportunity to improve, innovate, and inspire. So, go forth and build for the future.

Index

www.packtpub.com

Subscribe to our online digital library for full access to over 7,000 books and videos, as well as industry leading tools to help you plan your personal development and advance your career. For more information, please visit our website.

Why subscribe?

- Spend less time learning and more time coding with practical eBooks and Videos from over 4,000 industry professionals

- Improve your learning with Skill Plans built especially for you

- Get a free eBook or video every month

- Fully searchable for easy access to vital information

- Copy and paste, print, and bookmark content

Did you know that Packt offers eBook versions of every book published, with PDF and ePub files available? You can upgrade to the eBook version at packtpub.com and as a print book customer, you are entitled to a discount on the eBook copy. Get in touch with us at customercare@packtpub.com for more details.

At www.packtpub.com, you can also read a collection of free technical articles, sign up for a range of free newsletters, and receive exclusive discounts and offers on Packt books and eBooks.

Other Books You May Enjoy

If you enjoyed this book, you may be interested in these other books by Packt:

React Key Concepts

Maximilian Schwarzmüller

ISBN: 978-1-80323-450-2

- Build modern, user-friendly, and reactive web apps
- Create components and utilize props to pass data between them
- Handle events, perform state updates, and manage conditional content
- Apply styles dynamically and conditionally to create a modern UI
- Use advanced state management techniques such as React's context API
- Utilize React router to render different pages for different URLs
- Understand key best practices and optimization opportunities

React 18 Design Patterns and Best Practices - Fourth Edition

Carlos Santana Roldán

ISBN: 978-1-80323-310-9

- Get familiar with the new React 18 and Node 19 features
- Explore TypeScript's basic and advanced capabilities
- Make components communicate with each other by applying various patterns and techniques
- Dive into MonoRepo architecture
- Use server-side rendering to make applications load faster
- Write a comprehensive set of tests to create robust and maintainable code
- Build high-performing applications by styling and optimizing React components

Packt is searching for authors like you

If you're interested in becoming an author for Packt, please visit `authors.packtpub.com` and apply today. We have worked with thousands of developers and tech professionals, just like you, to help them share their insight with the global tech community. You can make a general application, apply for a specific hot topic that we are recruiting an author for, or submit your own idea.

Share Your Thoughts

Now you've finished *Building Micro Frontends with React 18*, we'd love to hear your thoughts! Scan the QR code below to go straight to the Amazon review page for this book and share your feedback or leave a review on the site that you purchased it from.

https://packt.link/r/1-804-61096-8

Your review is important to us and the tech community and will help us make sure we're delivering excellent quality content.

Download a free PDF copy of this book

Thanks for purchasing this book!

Do you like to read on the go but are unable to carry your print books everywhere?

Is your eBook purchase not compatible with the device of your choice?

Don't worry, now with every Packt book you get a DRM-free PDF version of that book at no cost.

Read anywhere, any place, on any device. Search, copy, and paste code from your favorite technical books directly into your application.

The perks don't stop there, you can get exclusive access to discounts, newsletters, and great free content in your inbox daily

Follow these simple steps to get the benefits:

1. Scan the QR code or visit the link below

https://packt.link/free-ebook/9781804610961

2. Submit your proof of purchase
3. That's it! We'll send your free PDF and other benefits to your email directly